ISAAC ASIMOV is undoubtedly America's foremost writer on science for the layman. An Associate Professor of Biochemistry at the Boston University School of Medicine, he has written well over a hundred books, as well as hundreds of articles in publications ranging from *Esquire* to Atomic Energy Commission pamphlets. Famed for his science fiction writing (his three-volume Hugo Award-winning THE FOUNDATION TRILOGY is available in individual Avon editions and as a one-volume Equinox edition), Dr. Asimov is equally acclaimed for such standards of science reportage as THE UNIVERSE, LIFE AND ENERGY, THE SOLAR SYSTEM AND BACK, ASIMOV'S BIOGRAPHICAL ENCYCLOPEDIA OF SCIENCE AND TECHNOLOGY, and ADDING A DIMENSION (all available in Avon editions). His non-science writings include the two-volume ASIMOV'S GUIDE TO SHAKESPEARE, ASIMOV'S ANNOTATED DON JUAN, and ASIMOV'S GUIDE TO THE BIBLE (available in a two-volume Avon edition). Born in Russia, Asimov came to this country with his parents at the age of three, and grew up in Brooklyn. In 1948 he received his Ph.D. in Chemistry at Columbia and then joined the faculty at Boston University, where he works today.

Other Avon books by
Isaac Asimov

ADDING A DIMENSION	36871	$1.50
ASIMOV'S BIOGRAPHICAL ENCYCLOPEDIA OF SCIENCE AND TECHNOLOGY	32775	6.95
ASIMOV'S GUIDE TO THE BIBLE, THE NEW TESTAMENT	24786	4.95
ASIMOV'S GUIDE TO THE BIBLE, THE OLD TESTAMENT	24794	4.95
FROM EARTH TO HEAVEN	29231	1.95
THE UNIVERSE: FROM FLAT EARTH TO QUASAR	27979	2.50
THE FOUNDATION TRILOGY	38026	5.95
FOUNDATION	38075	1.95
FOUNDATION AND EMPIRE	30627	1.95
SECOND FOUNDATION	38125	1.75
LIFE AND ENERGY	31666	1.95
THE NEUTRINO	38109	1.75
THE PLANET THAT WASN'T	35709	1.75
OF TIME, SPACE, AND OTHER THINGS	35584	1.75
THE SOLAR SYSTEM AND BACK	36012	1.50
VIEW FROM A HEIGHT	24547	1.25

ASIMOV ON PHYSICS

ISAAC ASIMOV

A DISCUS BOOK/PUBLISHED BY AVON BOOKS

AVON BOOKS
A division of
The Hearst Corporation
959 Eighth Avenue
New York, New York 10019

Published by arrangement with Doubleday and Company, Inc.
Library of Congress Catalog Card Number: 75-21205
ISBN: 0-380-41848-7

First Discus Printing, January, 1979

DISCUS TRADEMARK REG. U.S. PAT. OFF. AND IN
OTHER COUNTRIES, MARCA REGISTRADA, HECHO EN
U.S.A.

Printed in the U.S.A.

Dedicated to
the memory of P. Schuyler Miller, Will Jenkins
and Otto Binder

All essays in this volume originally appeared in *The Magazine of Fantasy and Science Fiction.* Individual essays were in the following issues:

Catching Up with Newton	December 1958
Of Capture and Escape	May 1959
The Ultimate Split of the Second	August 1959
The Height of Up	October 1959
C for *Celeritas*	November 1959
Thin Air	December 1959
Now Hear This!	December 1960
Order! Order!	February 1961
The Modern Demonology	January 1962
The Light Fantastic	August 1962
The Rigid Vacuum	April 1963
The Light That Failed	June 1963
A Piece of the Action	April 1964
First and Rearmost	October 1964
The Certainty of Uncertainty	April 1965
Behind the Teacher's Back	August 1965
The Land of Mu	October 1965

CONTENTS

INTRODUCTION

Back in 1959 I began writing a monthly science column for *The Magazine of Fantasy and Science Fiction.* I was given carte blanche as to subject matter, approach, style, and everything else, and I made full use of that. I have used the column to range through every science in an informal and very personal way so that of all the writing I do (and I do a great deal) nothing gives me so much pleasure as these monthly essays.

And as though that were not pleasure enough in itself, why, every time I complete seventeen essays, Doubleday & Company, Inc., puts them into a book and publishes them. Eleven books of my F & SF essays have been published by now, containing a total of 187 essays. A twelfth, of course, is on its way.

Few books, however, can be expected to sell indefinitely; at least, not well enough to be worth the investment of keeping them forever in print. The estimable gentlemen at Doubleday have therefore (with some reluctance, for they are fond of me and know how my lower lip tends to tremble on these occasions) allowed the first five of my books of essays to go out of print.

Out of *hardback* print, I hasten to say. All five of the books are flourishing in paperback, so that they are still available to the public. Nevertheless, there is a cachet about the hardback that I am reluctant to lose. It is the hardbacks that supply the libraries; and for those who really want a permanent addition to their large personal collections of Asimov books* there is nothing like a hardback.

My first impulse, then, was to ask the kind people at Doubleday to put the books back into print and gamble on a kind of second wind. This is done periodically in the case of my science fiction books, with success even when paper-

* Well, start one!

back editions are simultaneously available. But I could see that with my essays the case was different. My science fiction is ever fresh, but science essays do tend to get out of date, for the advance of science is inexorable.

And then I got to thinking . . .

I deliberately range widely over the various sciences both to satisfy my own restless interests and to give each member of my heterogeneous audience a chance to satisfy his own particular taste now and then. The result is that each collection of essays has some on astronomy, some on chemistry, some on physics, some on biology, and so on.

But what about the reader who is interested in science but is *particularly* interested in one particular branch? He has to read through all the articles in the book to find three or four that may be just up his alley.

Why not, then, go through the five out-of-print books, cull the articles on a particular branch of science and put them together in a more specialized volume. Doubleday agreed and I made up a collection of astronomical articles which appeared as *Asimov on Astronomy*. It was an experimental project, of course, and might have done very poorly. It did not; it did very well. The gentlemen at Doubleday rejoiced, and I got to work at once (grinning) and put together *Asimov on Chemistry*. And that did well too, so now here you have *Asimov on Physics*.

This volume, *Asimov on Physics*, has three articles from *Fact and Fancy*, five articles from *View from a Height*, three articles from *Adding a Dimension*, three articles from *Of Space and Time and Other Things*, and three articles from *From Earth to Heaven*.

The articles are arranged, not chronologically, but conceptually, and I have done my best to make them go from the most familiar to the most arcane and, where possible, have one lead into the next.

Aside from grouping the articles into a more homogeneous mass in an orderly arrangement, what more have I done? Well, the articles are anywhere from nine to sixteen years old and their age shows here and there. I feel rather pleased that the advance of science has not knocked out a single one of the articles here included, or even seriously dented any, but minor changes must be made, and I have made them.

In doing this, I have not revised the articles themselves since that would deprive you of the fun of seeing me eat my words now and then, or, anyway, chew them a little. So I have made the changes by adding footnotes here and there where something I said needed modification or where I was forced to make a change to avoid presenting misinformation in the course of the article.

In addition to that, my good friends at Doubleday decided to prepare these books on the individual branches of science in a consistent and more elaborate format than they have used for my ordinary essay collections and have added illustrations to which I have written captions that give information above and beyond what is in the essays themselves.

Finally, since the subject matter is so much more homogeneous than in my ordinary grab-bag essay collections, I have prepared an index which will, I hope, increase the usefulness of the book as reference.

So, although the individual essays are old, I hope you find the book new and useful just the same. And at least I have explained, in all honesty, exactly what I have done and why. The rest is up to you.

ISAAC ASIMOV
New York, October 1974

One
THIN AIR

Earth's atmosphere is now going through a period of scientific importance and prominence. To put it as colorfully (and yet as honestly) as possible, it is all the scientific rage.*

Once before in scientific history, Earth's atmosphere passed through a period of glamour. Let me tell you about that before I get to the current period.

To begin with, in ancient Greek times, air had all the dignity of an "element"; one of the abstract substances out of which the Universe was composed. According to the philosophers, culminating in Aristotle, the Universe was composed of "earth," "water," "air," and "fire" in four concentric shells, with earth at the bottom and fire at the top.

In modern terms, earth is equivalent to the lithosphere, the solid body of the planet itself. Water is the hydrosphere, or ocean; and air is the atmosphere. Fire is less obvious, being so high (according to Aristotle) as to be ordinarily imperceptible to human senses. However, storms roiled the sphere of fire and made fragments of it visible to us as lightning.

Even the sphere of fire reached only to the Moon. From the Moon outward, there was a fifth and heavenly "element," like none of those on our imperfect earth. Aristotle called it "ether." Medieval scholars called it "fifth element" but did so in Latin, so that the word came out "quintessence." That word survives today, meaning the purest and most essential part of anything.

Such a theory as to the structure of the universe presented early thinkers with few problems about the air. For instance, did the atmosphere ever come to an end as one

* This article appeared in December 1959, when the space age was only two years old and people had scarcely yet dared think beyond the upper reaches of our own atmosphere.

went upward? Sure it did. It came to an end at the point where the sphere of fire began.

You see, there was always *something* in the Aristotelian view. Just as earth gave way to water and water to air, with no gap between, so air gave way to fire and fire to ether. There was never *nothing*. As the ancient scholars said, "Nature abhors a vacuum."

Did the atmosphere weigh anything? Obviously not. You didn't feel any weight, did you? If a rock fell on you or a bucket's worth of water, you would feel the weight. But there's no feeling of weight to the air. Aristotle had an explanation for this. Earth and water had a natural tendency to move downward, as far as they could, toward the center of the universe (i.e., the center of the Earth).

Air, on the other hand, had a natural tendency to move upward, as anyone could plainly see. (Blow bubbles under water and *watch* them move upward—not that Aristotle would appeal to experiment, believing as he did that the light of reason was sufficient to penetrate the secrets of nature.) Since air lifted upward, it had no weight downward.

Aristotle flourished about 330 B.C. and his views were gospel for a long time.

Curtain falls. Two thousand years pass. Curtain rises.

Toward the end of his long and brilliant life, Galileo Galilei, the Italian scientist, grew interested in the fact that an ordinary water pump drawing water out of a well would not lift the water any higher than about 33 feet above the natural level. This no matter how vigorously and how pertinaciously the handle of the pump was operated.

Now people thought they knew how a pump worked. It was so designed that a tightly fitted piston moved upward within a cylinder, creating a vacuum. Since Nature abhorred a vacuum, water rushed upward to fill said vacuum and was trapped by a one-way valve. The process was repeated and repeated, more and more water rushed upward until it poured out the spout. Theoretically, this should go on forever, the water rising higher and higher as long as you worked the pump.

Then why didn't water rise more than 33 feet above its natural level? Galileo shook his head, and never did find an answer. He muttered gruffly that apparently Nature abhorred a vacuum only up to 33 feet and recommended that his pupil Evangelista Torricelli look into the matter.

In 1643, the year after Galileo's death, Torricelli did that. It occurred to him that what lifted the water isn't a fit of emotion on the part of Dame Nature, but the very unemotional weight of air pressing down on the water and forcing it upward into a vacuum (which would ordinarily be filled with a balancing weight of air). Water could not be forced higher than 33 feet because a column of water 33 feet high pressed down as hard as did the entire atmosphere, so that there was a balance. Even if a complete vacuum were pulled over the water, so that air down at well-water level pushed the column upward without any back air pressure, the weight of the water itself was enough to balance the total air pressure.

How to test this? If you could start with a column of water, say, 40 feet long, it should sink until the 33-foot level was reached. A 40-foot column of water would have more pressure at the bottom than the entire atmosphere. But how to handle 40 feet of water?

Well then, suppose you used a liquid denser than water. In that case, a shorter column would suffice to balance air's pressure. The densest liquid Torricelli knew of was mercury. This is about 13½ times as dense as water. Since 33 divided by 13½ is about 2½ feet, a column of 30 inches of mercury should balance the air pressure.

Torricelli filled a tube (closed at one end and a yard long) with mercury, put his thumb over the open end, and tipped it into an open container of mercury.† If the air had no weight, it would not press on the exposed mercury level in the container. All the mercury in the tube would therefore pour out.

The mercury in the tube started pouring out, to be sure, but only to the extent of a few inches. Fully 30 inches of mercury remained standing, supported by nothing, apparently. It was either magic or else Aristotle was wrong and air had weight. There was no choice; air had weight. Thus the first glamorous period of the atmosphere had begun.

Torricelli had invented the barometer, an instrument still

† I've decided it can't be that easy. It seems to me the weight of mercury would be more than the thumb would retain. Yet I haven't been able to find out just how Torricelli did it.

used today to measure air pressure as so many inches of mercury. Furthermore, in the upper part of the tube, in the few inches that had been vacated by the mercury, there was a vacuum, filled with nothing but some mercury vapor and darned little of that. It is called a Torricellian vacuum to this day and was the first decent man-made vacuum ever formed. It showed definitely that Nature didn't give a plugged nickel one way or the other for vacuums.

In 1650 Otto von Guericke, who happened to be mayor of the German city of Magdeburg, went a step further. He invented an air pump which could pump air out of an enclosure, forming a harder and harder vacuum; i.e. one that grew more and more vacuous.

Von Guericke then demonstrated the power of air pressure in a dramatic way. He had two metal hemispheres made which ended in flat rims that could be greased and stuck together. If this were done, the heavy hemispheres fell apart of themselves. There was nothing to hold them together.

But one of the hemispheres had a valved nozzle to which an air pump could be affixed. Von Guericke put the hemispheres together and pumped the air out of them, then closed the valve. Now the weight of the atmospheres was pressing the hemispheres together and there was no equivalent pressure within.

How strong was the air pressure? Well, publicity-wise von Guericke attached a team of horses to one hemisphere by a handle he had thoughtfully provided upon it and another team to the other hemisphere. With half the town of Magdeburg watching open-mouthed, he had the horses strain uselessly in opposite directions.

The thin air about us which "obviously" weighed nothing did indeed weigh plenty. And when that weight was put to use, two teams of horses couldn't counter it.

Von Guericke released the horses, opened the valve, and the hemispheres fell open by themselves. It was as dramatic an experiment as Galileo's supposed tossing of two balls of different mass of the Tower of Pisa, and what's more, von Guericke's experiment really happened. (They don't make mayors like that anymore.)

Since the atmosphere has weight, there could only be so much of it and no more. There could only be enough of it to allow a column of air (from sea level to the very tiptop),

The Bettmann Archive, Inc.

THE BAROMETER

There are few instruments as simple as the barometer. In essence, it is just a column of mercury held up by air pressure. Its mere existence, however, had astonishing consequences, for it showed that air pressure had a certain definite value and that there was only a certain finite amount of air above the surface of the Earth. The atmosphere, it became evident, could only be a few miles high.

When a barometer was taken up a mountain, the steady drop in its reading confirmed this and showed further that as one went higher, the air grew thinner, for the barometric reading dropped at a rate slower than it would if the atmosphere were of constant density with height. There was no question, though, that the atmosphere (except for negligible traces) stopped long before the Moon could be reached.

For the first time, human beings realized that space was largely vacuum. Until then, it had taken for granted that the entire space between heaven and earth was filled with air. Those who speculated on trips to the Moon before the days of the barometer (and even for some time afterward —before grasping the significance) assumed there would be air all the way. They imagined reaching the Moon by hitching a chariot to a team of large birds, for instance. Even as late as the 1830s, Poe wrote a tale (a not very serious one) in which his hero tried to reach the Moon by balloon.

The fact that space was vacuum placed severe constraints on spaceflight. There was nothing to "push against" so that all earthly systems for producing motion were useless. Only Newton's third law of motion would work. Part of an object would have to be hurled in one direction so that the rest would move in the other direction. The rocket worked by that principle and worked better in vacuum than in air, provided it carried its air supply with it. It was by rocket and the principle of the third law that the Moon was finally reached.

with a cross-sectional area of one square inch, to weigh 14.7 pounds. If the atmosphere were as dense all the way up as it is at sea level, a column just five miles high would have the necessary weight.

But of course, air isn't equally dense all the way up.

In the 1650s a British scientist, Robert Boyle, having read of von Guericke's experiments, set about to study the properties of air more thoroughly. He found it to be compressible.

That is, if he trapped a sample of air in the short closed half of a U-tube by pouring mercury into the long, open half, the trapped air contracted in volume (i.e. was compressed) until it had built up an internal pressure that balanced the head of mercury. As the mercury was added or removed the trapped air compressed and expanded like a spring. The English scientist, Robert Hooke, had just been reporting on the behavior of actual springs and since the trapped air behaved analogously, Boyle called it "the spring of the air."

If, now, Boyle poured additional mercury into the U-tube, the trapped air decreased further in volume until the internal pressure had increased to the point where the additional weight of mercury could be supported. Furthermore, Boyle made actual measurements and found that if the pressure on the trapped air was doubled, its volume was halved; if the pressure was tripled, the volume was reduced to one third and so on. (This is one way of stating what is now called Boyle's law.)

This was a remarkable discovery, for liquids and solids did not behave in this way. Boyle's work marks the beginning of the scientific study of the properties of gases which, in a hundred years, produced the atomic theory and revolutionized chemistry. This was just another consequence of this first glamorous period of the atmosphere.

Since air is compressible, the lowest regions of the atmosphere, which bear all the weight of all the air above must be most compressed. As one moves upward in the atmosphere, each successive sample of air at greater and greater heights has less atmosphere above it, is subjected to a smaller weight of air, and is therefore less compressed.

It follows that a given number of molecules takes up more room ten miles up than they do at sea level, and more room still twenty miles up and more room still thirty miles up and so on indefinitely. From this, it would seem that the atmosphere must also stretch up indefinitely. True, there's less and less of it as you go up, but that less and less is taking up more and more room.

In fact, it can be calculated that, if the atmosphere were at the sea-level average of temperature throughout its height, air pressure would be reduced tenfold for every twelve miles we travel upward. In other words, since the air pressure is 30 inches of mercury at sea level, it would be 3 inches of mercury at a height of 12 miles, 0.3 inches of mercury at 24 miles, 0.03 inches of mercury at 36 miles, and so on.

Even at a height of 108 miles, there would still be, by this accounting, 0.00000003 inches of mercury of pressure. This doesn't sound like much, but it means that six million tons of air would be included in the portion of the atmosphere higher than 100 miles above Earth's surface.

Of course, the atmosphere is *not* the same temperature throughout. It is the common experience of mankind that mountain slopes are always cooler than the valley below.

There is also no denying the fact that high mountains are perpetually snow-covered at the top, even through the summer and even in the tropics.

Presumably, then, the temperature of the atmosphere lowers with height and, it seemed likely, did so in a smooth fall all the way up. This spoiled the simple theory of decline of density with height but it didn't alter the fact that the atmosphere was remarkably high. Once astronomers started looking, they found ample evidence of that.

For instance, visible meteor trails have been placed (by triangulation) as high as 100 miles. That means that even at 100 miles, then, there is enough atmosphere to friction tiny bits of metals to incandescence.

Furthermore, aurora borealis (caused by the glowing of thin wisps of gas as the result of the bombardment with particles from outer space) has been detected as high as 600 miles.

However, how as one to get details on the upper atmosphere? Particularly one would want to know the exact way in which temperature and pressure fell off with height. As early as 1648 the French scientist, Blaise Pascal, had sent a friend†† up a mountain side with a barometer to check the fall of air pressure; but then, how high are mountains?

The highest mountains easily accessible to the Europeans of the seventeenth century were the Alps, the tallest peaks of which extend 3 miles into the air. Even the highest mountains of all, the Himalayas, only double that. And then, how could you be sure that the air 6 miles high in the Himalayas was the same as the air 6 miles high over the blank and level ocean.

No, anything in the atmosphere higher than, say, a mile was attainable only in restricted portions of the globe and then with great difficulty. And anything higher than 5 or 6 miles just wasn't attainable, period. No one would ever know. No one.

So the first glamorous period of the atmosphere came to an end.

Curtain falls. A century and a half passes. Curtain rises.

In 1782 two French brothers, Joseph Michel Montgolfier and Jacques Étienne Montgolfier, lit a fire under a large light bag with an opening underneath and allowed the

†† His brother-in-law actually.

heated air and smoke to fill it. The hot air, being lighter than the cold air, moved upward, just as an air bubble would move upward in water. The movement carried the bag with it, and the first balloon had been constructed.

Within a matter of months, hydrogen replaced hot air, gondolas were added, and first animals and then men went aloft. In the next few decades, aeronautics was an established craze—a full century before the Wright brothers.

Within a year of the first balloon, an American named John Jeffries went up in one, carrying a barometer and other instruments, plus provisions to collect air at various heights. The atmosphere, miles high, was thus suddenly and spectacularly made available to science and the second glamorous period had begun.

By 1804 the French scientist, Joseph Louis Gay-Lussac, had gone up nearly 4½ miles in a balloon, a height considerably greater than that of the highest peak of the Alps, and brought down air collected there.

It was, however, difficult to go much higher than that, because the aeronauts had the inconvenient habit of breathing. In 1874, three men went up 6 miles—the height of Mount Everest—but only one survived. In 1892 the practice of sending up unmanned (but instrumented) balloons was inaugurated.

The most important purpose of the early experiments was the measurement of the temperature at heights and by the 1890s some startling results showed up. The temperature did indeed drop steadily as one went upward, until at a height somewhat greater than that of Mount Everest, the temperature of −70° F. was reached. Then, for some miles higher, *there were no further temperature changes.*

The French meteorologist, Leon P. Teisserenc de Bort, one of the discovers of this fact, therefore divided the atmosphere into two layers. The lower layer, where there was temperature change, was characterized by rising and falling air currents that kept that region of the atmosphere churned up and produced clouds and all the changing weather phenomena with which we are familiar. This is the *troposphere* ("the sphere of change").

The height at which the temperature fall ceased was the *tropopause* ("end of change") and above it was the region of constant temperature, a place of no currents or churning, where the air lay quietly and (Teisserenc de Bort thought) in layers, with the lighter gases floating on top. Perhaps the earth's atmospheric supply of helium and hydrogen were to be found up there, floating on the denser gases below. He called this upper layer the *stratosphere* ("sphere of layers").

The tropopause is about ten miles above sea level at the equator and only five miles above at the poles. The stratosphere extends from the tropopause up to about sixteen miles. There, where the temperature starts changing again, is the *stratopause.*

About 75 per cent of the total air mass of the earth exists within the troposphere and another 13 per cent is in the stratosphere. Together, troposphere and stratosphere, with 98 per cent of the total air mass between them, make up the "lower atmosphere." But it is the 2 per cent above the stratopause, the "upper atmosphere," which gained particular prominence as the twentieth century wore on.

In the 1930s ballooning entered a new era. Balloons of polyethylene plastic were lighter, stronger, less permeable to gas than the old silken balloons (cheaper, too). They could reach heights of more than twenty miles. Sealed gondolas were used and the balloonists carried their own air supply with them.

In this way, manned balloons reached the stratosphere and beyond. Russian balloonists brought back samples of stratospheric air and no helium or hydrogen was present; just the usual oxygen and nitrogen. (We now know that the atmosphere is largely oxygen and nitrogen all the way up.) *

Airplanes with sealed cabins were flying the stratosphere, too, and toward the end of World War II, the *jet streams* were discovered. These were two strong air currents girdling the earth, and moving from east to west at 100 to 500 miles per hour at about tropopause heights, one in the North Temperate Zone and one in the South Temperate. Apparently they are of particular importance in weather forecasting, for they wriggle about quite a bit and the weather pattern follows their wriggling.

After World War II, rockets began going up and sending down data. The region above the stratosphere was more

* Not *all* the way up. Before the atmosphere thins out to the general density of interplanetary matter, it has a layer of helium from a height of 200 miles above the surface to 600 miles above and a layer of hydrogen beyond that. This excessively rarefied "heliosphere" and "protonosphere" were detected in rocket experiments conducted after this article appeared.

and more thoroughly explored. Thus it was found that from the stratopause to a height of about 35 miles, the temperature *rises*, reaching a high of −55° F. before dropping once more to −100° F. at a height of about 50 miles. Above that there is a large and steady rise to temperatures that are estimated to be about 2200° F. at a height of 300 miles and are probably higher still at greater heights.

The region of rising, then falling, temperature, from 16 to 50 miles is now called the *mesosphere* ("the middle sphere") and the region of minimum temperature that tops it is the *mesopause*. The mesosphere contains virtually all the mass of the upper atmosphere, about 2 per cent of the total. Above the mesopause, only a few thousandths of a per cent of the atmosphere remain.

These last wisps are, however, anything but insignificant, and they are divided into two regions. From 50 to 100 miles is the region where meteor trails are visible. This is the *thermosphere* ("sphere of heat" because of the rising temperatures) and is topped by the *thermopause* though that is *not* the "end of heat." Some authorities run the thermosphere up to 200 or even 300 miles.

Above the thermopause, is the region of the atmosphere which is too thin to heat meteors to incandescence but which can still support the aurora borealis. This is the *exosphere* ("outside sphere").†

There is no clear upper boundary of the exosphere. Actually, the exosphere just thins and fades into interplanetary space (which is *not*, of course, a complete vacuum). Some try to judge the "end of the atmosphere" by the manner in which the molecules of the air hit one another.

Here at sea level, molecules are crowded so closely together that any one molecule will only be able to travel a few millionths of an inch (on the average) before striking another. The air acts like a continuous medium, for that reason.

At a height of ten miles, the molecules have so thinned out that they may travel a ten-thousandth of an inch before colliding. At a height of 70 miles, they will travel a yard and a half and at 150 miles, 370 yards before colliding. At a height of several hundred miles, collisions become so rare that you can ignore them and the atmosphere begins to behave like a collection of independent particles.

† It is in the exosphere that helium and hydrogen are to be found.

(If you have ever been part of the New Year's Eve crowd in Times Square, and have also walked a lonely city street at 2 A.M., you have an intuitive notion of the difference between particles composing an apparently continuous medium and particles in isolation.)

The point where the atmosphere stops behaving like a continuous medium and begins to act like a collection of independent particles may be considered the *exopause*, the end of the atmosphere. This had been placed at heights varying from 600 to 1000 miles by different authorities.

The practical importance to us of the upper atmosphere is that it bears the brunt of the various bombardments from outer space, blunting them and shielding us.

For one thing there is the Sun's heat. The Sun emits photons with the energy one would expect of a body with the surface temperature of 10,000° F. These photons do not lose energy as they travel through space, and consequently strike the atmosphere in full force. Fortunately, the Sun radiates them in all directions and only a billionth or so are intercepted by our own planet.

Still, when one of the photons strikes a molecule at the edge of the atmosphere and is absorbed, that molecule may find itself possessed of a Sun-type temperature of 10,000° F. Only a small proportion of the molecules of Earth's atmosphere are so heated and slowly, by collision with other molecules below, the energy is shared so that the temperature drops to bearable levels as one descends.

(The high temperatures of the exosphere and thermosphere are an odd echo of the Aristotelian sphere of "fire." You may also be wondering how rockets can pass through the exosphere, if it has a temperature in the thousands of degrees, without being destroyed. There you run up against the difference between temperature and heat. The individual molecules have much energy, i.e. have a high temperature, but there are so few of them, that the total energy, i.e. heat, is negligible.)

Of course, the high temperature of the outermost atmosphere has its effects on the molecules that compose it. Oxygen and nitrogen molecules, shaken by this temperature and exposed to the bombardment of high energy particles besides, break up into individual atoms. (If the free atoms sink down to positions where less energy is available, they recombine, so no permanent damage is done.)

People have wondered whether ram-jets might not make use of these free atoms to navigate the exosphere. If enough could be gathered and compressed (and that is the hard part) the energy delivered per weight by their reunion to form molecules would be much higher than the energy delivered per weight by the combination of conventional fuels with oxygen, ozone, or fluorine.

Furthermore, the supply would be inexhaustible, since the atoms, once combined into molecules, would be expelled out the rear where the Sun's energy would promptly split them into atoms again. In effect, such a ram-jet would be running on solar energy, one tiny step removed.

The bombardment of particles from space also succeeds in damaging individual atoms or molecules, knocking off one or more planetary electrons, and leaving behind charged atom fragments called *ions*. Enough ions are formed in the exosphere to produce the glow called the auroras.

In the denser air of the thermosphere, there are more or less permanent layers of ions at different heights. These first made themselves known by the fact that they reflect certain radio waves. In 1902 Oliver Heaviside of England and Arthur Edwin Kennelly of the United States discovered (independently) the lowest of these layers, about 70 miles high. It is called the Kennelly-Heaviside layer in their honor.

Higher layers (at about 120 miles and 200 miles) were discovered in 1927 by the British physicist, Edward Victor Appleton, and these are called the Appleton layers. Because of these various layers of ions, the thermosphere is frequently called the *ionosphere*, and its upper boundary the *ionopause* (though that is not the "end of ions" any more than the "end of heat").

Nowadays, the layers have received objective letters. The Kennelly-Heaviside layer is the E layer, while the Appleton layers are the F_1 layer and F_2 layer. Between the F_1 layer and the E layer is the E region and below the E layer is the D region.

Yet lower in the atmosphere, down in the mesosphere, the ultraviolet of the Sun is still capable of inducing chemical reactions that do not ordinarily proceed spontaneously at sea level. It is possible to send chemicals up there and watch things happen. In the main, though, the important point is that something happens to a chemical already present there. Ordinary oxygen molecules of the mesosphere (made up of two oxygen atoms apiece) are converted into the more energetic ozone molecules (made up of three oxygen atoms apiece).

The ozone is continually changing back to oxygen while the forever incoming ultraviolet is continually forming more ozone. An equilibrium is reached and a permanent layer of ozone exists about 15 miles above the Earth's surface. This is fortunate for us since the maintenance of the ozone layer continually absorbs the Sun's hard ultraviolet which, if it were allowed to reach the Earth's surface unabsorbed, would produce seriously undesirable effects.

Because of the chemical reactions proceeding in the mesosphere it is sometimes called the *chemosphere* (and its upper boundary, the *chemopause*). As for the ozone layer itself, that is sometimes referred to as the *ozonosphere.*††

So there you have the steps. From Aristotle's undifferentiated "air" through one period of scientific glamour to Boyle's smoothly thinning atmosphere; then through another period of scientific glamour to the modern layers upon layers of air, with changing properties.

Next step (now begun): the investigation of *cis-Lunar space* (the space "this side of the Moon") which has already yielded the surprising knowledge of the existence of the Van Allen radiation belts—and what else?

Well, wait and see.*

†† In 1974, the ozonosphere made front-page headlines. There were reports that Freon used to push aerosol sprays out of cans might gradually drift up to the ozonosphere, release chlorine atoms which would deplete or destroy the ozone present with possibly dire consequences. A rather unsettling thought.

* The Van Allen belts are now called the "magnetosphere." High speed particles, shooting out of the Sun in all directions (the "solar wind"), are deflected by the magnetosphere as they approach the Earth, but drive it back, producing a sharp cut-off in the magnetosphere called the "magnetopause." The magnetopause is comparatively close to Earth on the Sun-side but is farther away on the sides and tails off in a long teardrop shape on Earth's night-side.

Two
NOW HEAR THIS!

The ancient Greeks weren't always wrong.

I am taking the trouble to say this strictly for my own good, for when I trace back the history of some scientific concept, I generally start with the Greeks, then go to great pains to show how their wrong guesses had to be slowly and carefully corrected by the great scientists of the sixteenth and seventeenth centuries, usually against the strenuous opposition of traditionalists. By the time I have done this on several dozen different occasions, I begin, as a matter of autohypnosis, to think that the only function served by the ancient philosophers was to put everyone on the wrong track.

And yet, not entirely so. In some respects we are still barely catching up to the Greeks. For instance, the Navy is now* studying dolphins and porpoises. (These are small-sized relatives of the whale, which differ among themselves most noticably in that dolphins have lips that protrude in a kind of beak while porpoises do not. I will, however, use the two terms interchangeably and without any attempt at consistency.)

In recent years, biologists have begun making observations concerning the great intelligence of these creatures.

For instance, dolphins never attack men. They may play games with them and pretend to bite, but they don't really. There are even three cases on record of the creatures guiding men, who have fallen overboard, back to shore. On the other hand, a dolphin, who had earlier played harmlessly with a man, promptly killed a barracuda placed in its tank with one snap of its jaws.

Now I'm not sure that it is a true sign of intelligence to play gently with men and kill barracudas, considering that

* This article appeared in December 1960. Since then, alas, the excitement over dolphins has subsided. That their intelligence is great is not in dispute but the hopes of establishing communication with them are fading.

man is much the more ferocious and dangerous creature. However, since man himself is acting as the judge of intelligence, there is no question but that he will give the dolphin high marks for behaving in a fashion he cannot help but approve.

Fortunately, there are more objective reasons for suspecting the existence of intelligence in dolphins. Although only five to eleven feet long, and therefore not very much more massive than men, they have a larger and more convoluted brain. It is not so much the size of the brain that counts as a measure of intelligence as the extent of its surface area, since upon that depends the quantity of gray matter. As, in the course of evolution, the brain surface increased faster than space for it was made available, it had to fold into convolutions. The extent and number of convolutions increases as one goes from opossum to cat to monkey to man, but it is the cetaceans (a general term for whales, dolphins and porpoises), and *not* man, that hold the record in this respect.

Well, then, is the intelligence of the dolphin, so clearly advertised by its brain, really a new discovery? I doubt it strongly. I think the Greeks anticipated us by a few millennia.

For instance, there is the old Greek tale of Arion, a lyre-player and singer in the employ of Periander of Corinth. Aron, having won numerous prizes at a music festival in Sicily, took passage back to Corinth. While en route the honest sailors on board saw no reason why they could not earn a bonus for themselves by the simple expedient of throwing Arion overboard and appropriating his prizes.

Being men of decision and action, they were about to do exactly that when Arion asked the privilege of singing one last song. This was granted by the sailors who, after all, were killing Arion only by way of business and not out of personal animosity.

Arion's sweet song attracted a school of dolphins, and when he jumped overboard at its end, one of the dolphins took him on his back and sped him back to Corinth faster than the ship itself could travel. When the sailors arrived in port, Arion was there to bear witness against them and they were executed. This story doesn't say whether they were allowed to sing a final song before being killed.

But why a dolphin? Surely if the Greeks were merely

composing a fantasy, a shark would have done as well, or a giant sea horse, or a merman, or a monstrous snail. Yet they chose a dolphin not only in the Arion tale but in several others. It seems clear to me that dolphins were chosen deliberately because Greek sailors had observed just those characteristics of the creatures that the Navy is now observing once more—their intelligence and (no other word will do) kindliness.

Oddly enough, the Greeks, or at least one Greek, was far ahead of his time in another matter which involved the dolphin.

Let me backtrack a little in order to explain. In ancient times, living creatures were classified into broad groups depending on characteristics that were as obvious as possible. For instance, anything that lived in the water permanently was a fish.

Nowadays, to be sure, we restrict the word fish to vertebrates which have scales and breathe by means of gills. Invertebrates such as clams, oysters, lobsters and crabs are *not* fish. However, the English language has never caught up with the modern subtleties of taxonomy. These non-fish are all lumped together as "shellfish." Even more primitive creatures receive the appellation, so that we speak of "starfish" and "jellyfish."

By modern definition, even a sea-dwelling vertebrate is not a fish if it lacks gills and scales, which means that whales and their smaller relatives are not fish. To the modern biologist, this seems obvious. The cetaceans are warm-blooded, breathe by means of lungs and, in many ways, show clearly that they are descended from land-dwelling creatures.

However, they have become so completely adapted to the sea that they have lost any visible trace of their hind limbs, transformed their front limbs into vaguely fishlike flippers, and developed a tail that is horizontal rather than vertical but otherwise again superficially fishlike. They have even streamlined themselves into a completely fishlike shape. For all these reasons, what is obvious to the biologist is not obvious to the general public, and popular speech insists on calling a whale a fish.

Thus, the song "It Ain't Necessarily So" from *Porgy and Bess* speaks of Jonah living in a whale and he is described as having "made his home in that fish's abdomen." Those of us who are sophisticated and know the difference between mammals and fish by the modern definition, can laugh good-humoredly at the charming simplicity of the characters in the play; but actually, the mistake is in reverse, and almost all of us are involved in it.

The Book of Jonah does *not*—I repeat, does *not*—mention any whale. Jonah 1:17 (in the King James version) reads: "Now the Lord had prepared a great fish to swallow up Jonah. And Jonah was in the belly of the fish three days and three nights." The creature is mentioned again in Jonah 2:1 and Jonah 2:10, each time as a fish.

It is only folk taxonomy that converted "great fish" into "whale."

And now I am ready for the Greeks again. The first to distinguish cetaceans from other sea creatures was Aristotle, back about 340 B.C. In a book called *Generation of Animals* he pointed out something that, considering the times, was a miracle of accurate observation; that is, that dolphins brought forth young alive, and that the young dolphins, when born, were attached to their mothers by an umbilical cord. Now, an umbilical cord implies that the embryo derives nourishment from the mother, continuously and directly, and not from a fixed food supply within an egg, as is the case with, for instance, certain snakes that bring forth their young alive. The umbilical cord is characteristic of the hairy, milk-yielding quadrupeds we call mammals, and of no other creatures.

For that reason, Aristotle classified the cetaceans with the mammals and not with the fish.

Frustratingly enough, where so many of Aristotle's wrong guesses and deductions were held to by ancient and medieval thinkers in a kind of death grip, this accurate observation which fits perfectly with our modern ways of thinking, was ignored. It wasn't until the nineteenth century that Aristotle's statements about the dolphin were finally confirmed.

One thing we have learned recently about porpoises, that the ancient Greeks probably did not know, concerns the noises they make. Now microphones under the sea have shown us that the ocean (surprisingly enough) is a noisy place, with shellfish clicking their claws and fish grunting weirdly. However, the cetaceans are the only creatures, aside from man, with brains complex enough to permit the

Courtesy National Marine Fisheries Service, National Oceanic and Atmospheric Administration

CETACEANS

One wonders how the cetaceans—whales, dolphins (here shown), and porpoises—managed to develop their complex brains, the only brains in the history of life to compare with the human. The brain must have developed on land, since the land environment is so much harsher than that of the sea that brain development is placed at a premium. Once a brainy land animal returns to the sea it retains its brains. Otters and seals are not less intelligent than their nearest land relatives.

But then otters and seals remain land animals in part. Only the cetaceans have broken away from land *completely* and yet still retain their brains in an environment which would seem to subject them to comparatively little stimulation. (Sharks do as well as cetaceans in the sea and sharks are awfully dumb.)

Granted that cetaceans have wonderful brains, what do they *do* with them? The primate brain is stimulated by all the environment information that enters by way of excellent eyes and extraordinary manipulative hands. Elephants have their strong and delicate trunk with which to explore the environment. Other relatively brainy animals have paws. The cetaceans have nothing—poor vision, no manipulative organs.

Everything about cetaceans is mysterious. We don't even know how they evolved. We have no record of any land animals we know to have been directly ancestral to whales. We don't know how the whale developed the ability to give birth at sea. (All other water-dwelling vertebrates come back to land to lay eggs or give birth to live young.)

Barring man, the cetaceans are among the most secure animals. The "killer whale" (actually the largest of the dolphins) need experience no fear. No other species preys on it or is a danger to it. Man himself, even when armed, need fear many of the larger predators, but not the killer whale. It, more nearly than any other species, is at the very top of the food chain.

delicate muscular movements that can produce sounds in wide variety. Porpoises do, in fact, whistle and rasp and grunt and creak in all sorts of ways. What's more, they have very well developed inner ears and can hear perfectly well all the sounds they make.

Surely, the thought should occur to us at this point that if porpoises are so all-fired smart and make all those noises, then perhaps they are *talking*. After all, we might argue, what else is such a battery of sound-formation good for?

Unfortunately, there is something aside from communication that sound is good for, and this noncommunicative use definitely exists in the case of the porpoise.

To explain this, let me go a bit into the nature of sound. Here, once more, we have a case where the Greeks got started on the right track (just about the only branch of physics in which they did).

Thus, Pythagoras of Samos observed, about 500 B.C., that

if two lyre strings were plucked, the shorter string emitted the higher-pitched sound. He, or some of his followers, may also have observed that both strings vibrated while sounding and that the shorter string vibrated the faster. At any rate, by 400 B.C., a philosopher of the Pythagorean school, Archytas of Tarentum, stated that sound was produced by the striking together of bodies, swift motion producing high pitch and slow motion producing low pitch.

Then Aristotle came along and specifically included air among the bodies which produced sound when struck. He stated further that one part of the air struck the next so that sound was propagated through it until it reached the ear. Without a medium such as air or water, Aristotle pointed out, man would not hear sound. And here he was correct again.

By the end of ancient times, Boethius, the last of the Roman philosophers, writing about 500 A.D., was comparing sound waves to water waves. (Actually, sound waves are longitudinal, while water waves are transverse—a distinction I won't go into further here—but the analogy is a good one in many ways.)

Long after it was accepted that sound was a wave motion, the nature of light remained in sharp dispute. During the great birth of modern science in the seventeenth century, one group, following Christian Huyghens, believed light to be a wave phenomenon, too, like sound. A larger group, however, following Isaac Newton, believed it to be a stream of very small, very fast particles.

The reason why the Newtonian corpuscular theory of light held sway for a century and more was not only because of Newton's great prestige, but because of a line of argument that can be presented most simply as follows:

Water waves (the most familiar of all wave motions) bend around obstacles. Float a stick on water in the path of an outspreading ripple and the smooth circular arc of the ripple will be disturbed and bent into a more complicated pattern, but it will not be stopped or reflected. There will be no ripple-free "shadow" cast by the stick.

Sound also is not stopped by obstacles but bends around them. After all, you can hear a friend quite distinctly if he is calling to you from around a corner or from behind a tall fence.

Light, on the other hand, does not bend around obstacles,

but is absorbed or reflected by them, and you can see by this reflection. Furthermore, the obstacle produces a light-free area behind it, a shadow with sharp edges that proves the light rays to be traveling in absolutely straight lines. This is so different from the behavior of water waves or sound waves that it seemed clear that light could not be waves but must be particles which would travel in straight lines and would not bend around them.

Then, in 1801, an English physicist, Thomas Young, allowed a narrow beam of light to pass through two closely spaced holes. The two beams that resulted broadened and overlapped on a screen behind. But instead of a simple reinforcement of light and a consequent brightening at the region of overlap, a series of bands or fringes of light were produced, separated by bands of darkness.

Now how could two beams of light unite to give regions of darkness? There seemed no way of explaining this if light consisted of particles. However, if they consisted of waves, there would be regions where the waves of the two light beams would be in phase (that is, moving up and down together) and therefore yield a region of light brighter than either could produce separately. There would also be regions where the two waves would be out of phase (one moving up while the other moved down) so that the two would cancel and produce darkness. This phenomenon of "interference" could be duplicated in water waves to produce exactly the effect observed by Young in the case of light.

That established the wave nature of light at once. (Actually, the modern view of light has it possessing both wave and particle aspects, but we needn't bother about that here.)

Furthermore, from the width of the interference bands and the distance between the two holes through which the beams issued, it proved possible to calculate the wave length of the light waves. The shortest wave lengths (those of violet light) were as short as 0.000039 centimeters; while the longest (those of red light) attained a length of 0.000075 centimeters.

Then, in 1818, the French physicist Augustin Jean Fresnel worked out the mathematics of wave motion and showed that waves would only travel about obstructions that were small compared to the wave length. A stick does not stop a water wave, but a long spit of land will; even in a storm, a region of the sea protected from open ocean by such a spit will remain relatively calm—in a wave-free "shadow." In fact, that's the whole principle underlying the value of harbors.

Conversely, objects visible to the naked eye, however small they may be, are large compared to the wave lengths of light, and that is why light waves don't bend around them, but are reflected and cast sharp shadows instead. If the obstructions are only made small enough, light waves *will* bend around them (a phenomenon called "diffraction"); this, Fresnel was able actually to demonstrate.

Now, then, back to sound. The wave lengths of sound must be much longer than those of light, because sound is diffracted by obstacles that stop light cold. (Nevertheless, although a tree will not reflect ordinary sound waves, mountains will, thus producing echoes, and sounds in large rooms will reverberate as sound waves bounce off the walls.)

The exact wave length of a particular sound wave can be obtained by dividing the velocity of sound by the frequency (that is, the number of times the sound source vibrates per second).

As for the velocity of sound, even primitive people must have known that sound has some finite speed, because at quite moderate distances you can see a woodcutter swing his ax and hit the tree, and then only after a moment or two will you hear the "thunk." Assuming that light travels at infinite velocity (and, compared with sound, it virtually does), all you would have to do to determine the velocity of sound would be to measure the time interval between seeing and hearing over a known distance.

The difficulty of timing that interval was too much for scholars until modern times. It wasn't until 1656 that the pendulum clock was invented (by Huyghens, by the way, who originated the wave theory of light); only then did it become possible to measure intervals of less than an hour with reasonable precision.

In 1738, French scientists set up cannons on hills some seventeen miles apart. They fired a cannon on one hill and timed the interval between flash and sound on the other; then they fired the cannon on the second hill and timed the interval on the first. (Doing it both ways and taking the average cancels the effect of the wind.) Thus, the velocity of sound was first measured. The value accepted today is 331 meters per second at 0° C., or 740 miles an hour.

The velocity of sound depends on the elasticity of air; that is, on the natural speed with which air molecules can bounce back and forth. The elasticity increases with temperature and so, therefore, does the velocity of sound, the increase coming to roughly half a meter per second for each Centigrade degree rise in temperature.

Middle C on the piano has a frequency of 264 vibrations per second; therefore its wave length is $^{331}/_{264}$ or 1.25 meters. Frequency goes up with higher pitch (as the Pythagoreans first discovered) and wave length therefore comes down. As pitch goes lower, frequency goes down and wave length up.

The lowest note on the piano has a frequency of 27.5 vibrations per second, while the highest note has a frequency of 4,224 vibrations per second. The wave length of the former is therefore $^{331}/_{27.5}$ or 12 meters, and of the latter is $^{331}/_{4,224}$ or 0.076 meters (which is equivalent to 7.6 centimeters).

Even the wide range of the piano doesn't represent the extremes of the ear's versatility. The normal human ear can hear frequencies as low as 15 vibrations per second and as high as 15,000 per second in adults and even 20,000 per second in children. This is an extreme span of over ten octaves (each octave representing a doubling of frequency) as compared with the single octave of light to which the eye is sensitive. In terms of wave length the human ear can make out a range, at the extreme, of from 22 meters down to less than 2 centimeters.

Even the highest-pitched sound we can hear, however, has a wave length about 20,000 times as long as that of red light, so that we have every reason to expect that sound and light should behave quite differently with regard to obstructions.

Still, the shorter the wave length (that is, the higher pitched the sound), the more efficiently an obstacle of a particular size ought to stop and reflect a sound wave. A tree should be able to reflect a 2-centimeter sound wave, while it would have no effect at all on a 22-meter sound wave.

Why not, then, progress still further down the wave-length scale and use sounds so high-pitched that they pass the limits of audibility? (These are "ultrasonic"—"beyond sound"—frequencies.) The existence of such inaudible sound is easily demonstrated even without man-made detectors. Whistles can be made which yield ultrasonic sound when they are blown. We hear nothing; but dogs, with ears capable of detecting sounds of higher frequency than ours can, come running.

The production of ultrasonic sounds in quantity first became possible as a result of a discovery made in 1880 by two brothers, Pierre and Jacques Curie. (Pierre Curie, a brilliant scientist, happened to marry a still more brilliant one—Marie, the famous Madame Curie—and is the only great scientist in history who is consistently identified as the husband of someone else.)

The Curie brothers found that quartz crystals, if cut in the proper manner, would, when slightly compressed as a result of very high pressure, develop small electric charges on their opposite faces. This is called "piezoelectricity" from a Greek word meaning "pressure." They also discovered the reverse phenomenon: if a difference in voltage is set up in metal plates held against opposite faces of the crystal, a small compression is induced in the crystal. From this it follows that if voltage is applied and removed rapidly, the crystal will expand and contract as rapidly, to produce a sound wave of that particular frequency. It will set up an ultrasonic beam if the vibration is rapid enough.

After the radio tube was developed, the production of voltages oscillating at ultrasonic frequencies became quite practical; the French physicist Paul Langevin succeeded in producing strong ultrasonic beams in 1917. World War I was going on, and he at once attempted to put to use the fact that sounds of such short wave length could be more efficiently reflected by relatively small obstacles. He used them to detect submarines under water. From the time lapse between emission of the ultrasonic pulse and detection of the echo and from the velocity of sound in water (which is over four times as great as the velocity in air, because of water's greater elasticity), the distance of the obstruction can be determined.

After World War I, this principle was put to peacetime use in detecting schools of fish and hidden icebergs, in determining the depths of the ocean and the conformation of the sea bottom, and in other ways. It went to war again in World War II, and received the name of "sonar," the abbreviation of "*s*ound *n*avigation *a*nd *r*anging."

But it would seem that sonar is one field in which mankind has been anticipated by other species of creatures by many millions of years.

The bat, for instance, is a clever flier, fluttering and flitting in an erratic course. (The original meaning of the word "bat" is "to flutter rapidly," as when you "bat your eyes," and an alternative name for the creature is "flitter-mouse.") In its wobbly course, the bat catches tiny insects with precision and evades small obstructions like twigs with ease. Considering that it flies about at twilight, it is amazing that it can do so.

In 1793, the Italian scientist Lazzaro Spallanzani found that bats could catch food and avoid obstacles even in pitch darkness and even if blinded. However, they lost this ability if they were deafened.

In the early 1940s, an American physicist, G. W. Pierce, developed a device that could pick up particularly faint ultrasonic beams; and it then turned out that bats were constantly emitting not only the faint squeaks that human ears could pick up, but inaudible ultrasonic sounds with frequencies as high as 150,000 vibrations per second, and wave lengths, consequently, as low as 2 millimeters.

Such short wave lengths are stopped reasonably well by insects and twigs. They are reflected, and the bats pick up the echoes between squeaks and guide themselves accordingly.

And this is exactly what porpoises and dolphins do also, in detecting fish rather than insects. Their larger prey makes it unnecessary for them to drive frequency so high and wave length so low. They do make use of the ultrasonic range, but accompanying that are sounds well into the audible range, too, usually described as "creaking."

Experiments at Woods Hole, Massachusetts, in 1955, showed that porpoises could pick up food fragments as small as 15 centimeters (6 inches) wide, even under conditions of complete darkness, provided they were creaking as they swam. If they were not creaking, they did not spot the fish. (One of the reasons the Navy is interested in these creatures, by the way, is that they hope to improve sonar technique by studying porpoise technique.)

This, then, is the noncommunicative use of sound referred to earlier in the article. We can speculate that perhaps the reason why sea life is so noisy is just because of the necessity of finding food and avoiding enemies in an environment where light is so limited and the sense of sight consequently so much less useful than it is on land.

But now let's ask a second question. Even if we grant that sound was first developed for purposes of sonar, a relatively simple scheme of sound-production should suffice (as in the bats). When sound-production is as elaborate as in porpoises, isn't it fair to consider it conceivable that a secondary use involving the elaboration may have developed?

To feel our way toward an answer to that question, let's consider man and certain experiments at Cornell in the early 1940s, which used blind people as well as normally sighted people who had been blindfolded.

These were made to walk down a long hall toward a fiberboard screen which might be in any position along the hall or might not be there at all, and they were to stop as soon as they were convinced the screen was just before them.

They did very well, spotting the screen almost every time some seven feet before they got to it. Most of the subjects were quite emphatic that they could somehow "feel" the approaching partition against their faces. When their heads were veiled so that the drapery would absorb any waves in the air that might be applying pressure to delicate hairs on the face, this was not found to interfere particularly with the ability to sense the screen.

However, when the ears of the subjects were efficiently plugged, the ability was lost at once. Apparently, the small echoes of footsteps or of other incidental noises gave the barrier away and, without knowing it, the men, both blind and blindfolded, were making use of the echo location principle.

The fact that human beings make use of sonar and that this, perhaps represents the original use of the sounds we make, has not prevented us from developing sound communication secondarily to the point where it is now the prime function of our vocal cords. It is not inconceivable, then, that porpoises, with as good a brain as our own and as good ears and as good sound-making equipment (or perhaps better in each case), might not also have developed speech.

Frankly, I wish most earnestly that they have. There are a few problems mankind has that I think might be solved if we could only talk them over with some creatures who could approach matters with a fresh and objective viewpoint.†

† And in the last fifteen years those problems have gotten a lot worse and we still have no one to talk to.

Three
CATCHING UP WITH NEWTON

It is very irritating that, in this modern era of missiles and satellites, there are so many newsmen who haven't caught up with Newton yet. They speak with appalling glibness about the weightlessness experienced by a spaceman once he has climbed "beyond the reach of gravity." Presumably they have the impression that there is a boundary line near the top of the atmosphere or thereabouts, beyond which there is suddenly no gravity—and that is the very thing Newton's theory disallows.

Isaac Newton was the first to formulate the Law of Universal Gravitation. Note the adjective "Universal," which is the important word. Newton did *not* discover that apples fell to the ground when they broke loose from the tree; that was common knowledge. What he did demonstrate was that the Moon's path around the Earth could be explained by supposing that the Moon was in the grip of the same force that tugged at the apple.

His great suggestion was that every piece of matter in the Universe attracted every other piece of matter, and that the quantity of this force could be expressed in a simple formula.

The force of attraction (f) between any two bodies, said Newton, is proportional to the product of the masses (m_1 and m_2) of the bodies and inversely proportional to the square of the distance (d) between their centers. By introducing a proportionality constant (G), we can set up an equation representing the above statement symbolically:

$$f = Gm_1m_2/d^2 \qquad \text{(Equation 1)}$$

The most recent and presumably most accurate value obtained for G (in 1928, at the Bureau of Standards) is 6.670×10^{-8} dyne cm^2/sec^2. This means that if two 1-gram spherical masses are placed exactly 1 centimeter apart (center to center), the attraction between them is 6.670×10^{-8} dynes.*

This shows gravity to be a relatively weak force as compared with electrical and magnetic attractions, for instance. One dyne of force is equivalent, roughly, to 1 milligram of weight. If the two 1-gram spheres were the only matter in the universe, therefore, each would weigh, under the gravitational attraction of the other at the distance indicated, only 0.000000066 milligrams (or about two-trillionths of an ounce). However, when masses as large as the Earth are concerned, even a weak force like gravity becomes tremendous.

Of course, we don't have to use dynes or any other fancy units to understand the essentials of gravity. Suppose, for instance, that the two masses between which we are trying to measure gravitational attraction are a spaceship and the planet Earth. The mass of the spaceship we can set equal to 1 (one what? one spaceship-mass). The mass of the Earth we can also set equal to 1, by using different units—one Earth-mass, this time.

The distance between the center of the Earth and the center of the spaceship, which we will suppose to be resting on the Earth's surface, is just about 3950 miles. We can make this value also 1 by calling that number of miles 1 Earth-radius.

Notice, now, that in using Newton's equation, it is necessary to take distances from center to center. In other words, the important point is not how far the spaceship is from the surface of the Earth, but how far from its center.

It is one of Newton's great accomplishments, you see, that he was able to demonstrate that spheres of uniform density attract each other as though all their mass were concentrated at the central point. To be sure, actual heavenly bodies are not uniformly dense, but Newton also showed this central-point business to be true for spheres which consisted of a series of layers (like an onion) each of which was uniform in density, though the density might vary from layer to layer. This modified situation *does* hold true for actual heavenly bodies.

But back to Earth and spaceship. Now that we have chosen convenient units for masses and distances, it is only

* As of 1974, the best value for G is 6.6720×10^{-8} dynes which is 0.03 per cent higher than the 1928 value.

necessary to make the gravitational constant also 1 (one constant-value) and Equation 1 becomes:

$$f = 1 \times 1 \times 1/1^2 \qquad \text{(Equation 2)}$$

Therefore, as the result of our shrewd unit choices, it turns out that the force of attraction between Earth and spaceship is exactly 1.

So far so good, but this is for the spaceship resting on Earth's surface. What if it were not on Earth's surface but 3950 miles straight up?

By changing the spaceship's position, we are not altering its mass, or Earth's mass or the gravitational constant. Each of these can remain 1. The only thing that is being altered is the distance between the center of the spaceship and the center of the Earth, so distance is all we need concern ourselves with and Equation 2 becomes:

$$f = 1/d^2 \qquad \text{(Equation 3)}$$

Now, then, if the spaceship is 3950 miles above the Earth's surface, its distance from the center of the Earth is 3950 miles plus 3950 miles or 2 Earth-radii. (We can use any units we want but, once having chosen them, we must stick with them. Such are the ethics of the situation.)

At 3950 miles above the Earth's surface, then, the force of attraction between Earth and the spaceship, using Equation 3, is $1/2^2$ or 0.25.

Gravitational attraction is usually measured by weighing an object. Consequently we can say that whatever the weight of the spaceship on the surface of the Earth, it weighs (i.e. is attracted by the Earth) only ¼ as much 3950 miles above the surface.

By the same reasoning we could show that this would hold for any object other than the spaceship. The gravitational attraction of the Earth for anything at all drops to a quarter of its value as that "anything at all" is moved from Earth's surface to a height 3950 miles above its surface.

Equation 3, will also give us the force between the Earth and the spaceship (or any other object) for any height above the surface. Some figures, so obtained, are shown in Table 1.

As you see, gravitational force starts dropping off at

TABLE 1 Earth's gravitational force in relation to distance

Distance to surface of earth	*Distance to center of earth*		*Gravitational attraction*
IN MILES	IN MILES	IN EARTH-RADII	
0 (sea-level)	3,950	1.000	1.000
50 (top of the stratosphere)	4,000	1.012	0.975
150	4,100	1.040	0.924
250	4,200	1.063	0.884
1,000	4,950	1.253	0.636
2,000	5,950	1.506	0.442
4,000	7,950	2.015	0.247
10,000	13,950	3.53	0.081
20,000	23,950	6.06	0.027
50,000	53,950	13.62	0.0054
100,000	103,950	26.3	0.0014
250,000 (moon-apogee)	253,950	64.2	0.00024
25,000,000 (closest approach of Venus)	25,003,950	6300	0.000000025

once. Even at low satellite-heights, so to speak, it varies from ⅔ to 9/10 of what it is at the planet's surface. Or, to get really petty about it: if you weigh 150 pounds and are suddenly transported to the top of Mount Everest from your sea-level home, you would find gravity weakened enough to make your weight 149½ pounds.

Nevertheless, Earth's gravitational force does not drop to zero, no matter what the distance. No matter how large you make d in Equation 3, f is never zero. If you go back to Equation 1, you would see that this is also true for the attraction between any two bodies, however small, with masses greater than zero. In other words, the gravitational influence of every body, however small, is exerted through all of space.

Nor does the force very quickly become negligible when large bodies are involved. The gravitational force between Earth and Venus, at closest approach is only 0.000000025 that of what it would be if the two planets were in contact. Nevertheless, the force attracting the Earth and Venus, even at a distance of 25,000,000 miles is still equal to 130 trillion tons.

So much for spacemen getting "beyond the reach of gravity."

The word "Universal" in Newton's law wouldn't be worth much, if we don't apply the equation to other bodies. We can start by supposing the spaceship to be resting on the surface of the Moon.

To begin with, the spaceship has the same mass (i.e. the quantity of matter contained in its substance) as on Earth and we agreed to let that mass (m_1) equal 1. The constant G never varies and we agreed to let that equal 1, also. Equation 1 therefore becomes:

$$f = m_2/d^2 \qquad \text{(Equation 4)}$$

where m_2 is the mass of the Moon and d is the distance from the center of the spaceship to the center of the Moon. Since the spaceship is on the Moon's surface, d is equal to the radius of the Moon.

We've defined our unit for m_2 as "Earth-masses" and for d as "Earth-radii," and we will stick to that. The Moon is only 0.0123 (about 1/81) as massive as the Earth and its radius is only 0.273 (a little over ¼) that of the Earth.

The Moon's mass is therefore 0.0123 "Earth-masses"

and its radius 0.273 "Earth-radii" so that Equation 2 becomes:

$$f = 0.0123/0.273^2 = 0.164 \quad \text{(Equation 5)}$$

This means that whatever the spaceship weighs on the surface of the Earth as the result of the force of Earth's gravitational attraction, it weighs 0.164 times that (roughly ⅙) on the surface of the Moon, as the result of the Moon's (lesser) gravitational attraction. By the same reasoning, this ratio of weight would hold true for any object at all.

Given the mass and radius of any body, the value of the surface gravity of that body can be calculated in the same way. The surface gravity of various bodies in the Solar System is presented in Table 2 by way of example.

TABLE 2 Some Surface Gravities in the Solar System

Astronomical Body	*Mass* (IN EARTH-MASSES)	*Radius* (IN EARTH-RADII)	*Surface gravity*
Jupiter (pole)	318.	10.5	2.88
Jupiter (equator)	318.	11.2	2.54
Neptune	17.3	3.4	1.50
Saturn (pole)	95.2	8.5	1.32
Saturn (equator)	95.2	9.5	1.05
Uranus	14.5	3.7	1.05
Earth	1.0	1.0	1.00
Venus	0.82	0.96	0.89
Mars	0.11	0.525	0.40
Mercury	0.054	0.380	0.27
Ganymede	0.026	0.395	0.17
Moon	0.0123	0.273	0.16

Notice that Jupiter and Saturn are not perfect spheres. Both are noticeably flattened at the poles. Saturn is the least spherical of the planets, there being a 12 per cent difference between the polar radius and the equatorial radius. For Jupiter, there is a 7.5 per cent difference. In both cases, since *d* varies with latitude, so does surface gravity, being least at the equator and highest at the pole. (The equatorial gravity is further decreased by the centrifugal force of the planet's spin, but I've ignored that here. Enough is enough.)

The fact that Saturn, which is so much more massive than Earth, has a surface gravity only slightly higher is not mysterious. Saturn is only ⅛ as dense as Earth and is correspondingly more voluminous than it would be if it were made of Earth-type material. The effect of the abnormally large radius for Saturn's mass (as compared with Earth) is to lower the surface gravity because of increased distance between Saturn's center and an object on its surface by just about as much as Saturn's increased mass (over Earth) raises it.

Surface gravities of Saturn and Earth may be approximately equal but this is illusory, in a way. Look at it this way—

A spaceship on a planetary surface is at varying distance from that planet's center, since planets come in different sizes. Suppose, though, that a spaceship is 230,000 miles from Earth's center at one time and 230,000 miles from Saturn's center at another.

When it is 230,000 miles from Earth's center it is about 226,000 miles above its surface. At 230,000 miles from Saturn's center, it is only 192,000 miles above its surface, Saturn being the larger body. However, in considering gravitational force, as I have pointed out, it is distance from the center that counts.

In such a case, with d equal in the two situations, only m_2 (see Equation 4) remains to vary the result. Earth's mass is, of course, equal to 1 "Earth-mass." Saturn's mass is 95.2 "Earth-masses." Therefore, the gravitational force gripping the spaceship in the neighborhood of Saturn is always 95.2 times that gripping it at an equal distance from Earth.

This can be shown in the behavior of two satellites that happen to be at this distance from Earth and Saturn. The Moon is at an average distance of 239,000 miles from Earth's center, while Saturn's satellite Dione is about 230,000 miles from Saturn's center. Each travels just about 1,500,000 miles in completing its circuit about its primary.

The greater the force of gravitational attraction upon a satellite, the faster must that satellite move to work up enough centrifugal force to keep in its orbit against its planet's pull. The Moon can manage this by traveling at a rate of 2200 miles an hour and completing its revolution in a leisurely 27.32 days. Dione, however, must race along

The Granger Collection

ROCKETS

The rockets that have made spaceflight possible are an advance that, more than any other technological victory of the twentieth century, was grounded in science fiction.

The first indication that rocket action might be a method for penetrating beyond the bonds of Earth to other worlds came in the 1650s in a science fiction story by Cyrano de Bergerac. Since then, the notion of spaceflight has been kept alive through centuries in which it remained impossible in reality, by the dreams of science fiction writers. Edgar Allan Poe, Jules Verne, and H. G. Wells all wrote of flights to the Moon, though none of them used rockets. Poe used a balloon, Verne a cannon, and Wells anti-gravity.

Serious scientists who worked out the rocket principle —Tsiolkovsky in Russia, Oberth in Rumania, Goddard in the United States—were influenced by science fiction. Tsiolkovsky even wrote some.

Goddard was the first actually to build rockets that might, in principle, penetrate beyond the Earth's atmosphere—carrying its own liquid fuel and liquid oxygen. The first such rocket rose into the air in 1926.

In Germany, Willy Ley and Wernher von Braun (both science fiction readers) took up the matter too. Willy Ley left Germany when Hitler came to power, but Von Braun remained and, with government help, finally developed the V-2 rocket that might conceivably have won the war for Germany, had it come along a few years sooner.

After the war both the United States and the Soviet Union developed the rocket into powerful tools for space exploration. The one pictured here lifted Gemini IV into orbit on June 3, 1965.

at just ten times that speed to stay in orbit. Its period of revolution is only 2.74 days.

That, and not the surface gravity figures, is a measure of the force a spaceship would be fighting if it were maneuvering in the neighborhood of Saturn.

Nevertheless, however great the gravitational force exerted by a planet, and however close to it a spaceship may be, it remains possible for the spaceship (and the people on it) to be weightless. And this does *not* mean that the force of gravity has been suspended.

Gravity is a force and a force is defined as something that can accelerate a mass. That, so to speak, is gravity's main job. It is what it is doing constantly all over the Universe.

We ourselves happen to be most used to gravitational force in its manifestation as the sensation of weight. Actually this type of manifestation occurs only in a special case: where a body is prevented from responding to gravitational force by accelerated motion. (Accelerated motion, by the way, is motion that is continually changing either in velocity or in direction or both.)

The most common way in which accelerated motion can be prevented is by having the two bodies between which the gravitational force exists (i.e. a spaceship and Earth) in contact so that neither can move with respect to the other under the pull of gravitational force alone. You and I are almost always in contact with Earth and it is for that reason that we learn to think of gravity as primarily concerned with weight.

Yet we live with the acceleration too. Hold a book at arm level and let go. At once gravitational force expresses itself in terms of acceleration. The book accelerates in the direction of Earth's center and keeps on until the surface of the planet intercepts it and it can move no more.

The Moon, as it moves about the Earth, is undergoing accelerated motion since, moving in an ellipse as it does, it is continually changing direction, turning a full 360° in 27.32 days. (It also continually changes velocity to a comparatively minor extent.) Dione, under the whip of a stronger gravitational force, is more strongly accelerated, changing direction more quickly and turning 360 degrees, as I have said, in only 2.74 days.

Whenever a body like a book or a satellite is responding to gravitational force by unrestricted accelerated motion, it is said to be in "free fall." The word "unrestricted" in the previous sentence is a bow in the direction of air resistance. A book falling from your hand ought to be moving through a vacuum to be in true free fall.

An object moving in response to gravitational force, with another constant (i.e. non-accelerated) motion superimposed, is still in free fall. A missile, with its charge expended, moving in a direction more or less opposed to that induced by gravity; or a satellite (artificial variety), with its rocket stages gone, and with a component motion perpendicular to that imposed by gravity—both are still in free fall.

An object which is competely in free fall is responding to gravity all it can; it has no response left over, so to speak, to be manifested as weight. An object in free fall is therefore weightless. A Cosmonaut orbiting the Earth in a satellite remains weightless as long as he stays in orbit. Gherman Titov remained weightless in this manner for a full day.† For that matter, if the cable of an elevator broke and it fell freely with unfortunate you inside, you would be as weightless for a few seconds (barring air-resistance effects) as any man in orbit in outer space.

If you were falling at an acceleration *greater* than that imposed by gravity (as in an airplane power dive) you would feel "negative weight." Within such a power-driving plane, you would fall upward at increasing speed (relative to the plane) unless you were strapped into your seat. This is one kind of "anti-gravity" which may not be useful but which is at least completely valid.

In calculating the force of gravity at various distances from Earth and on the surface of various planets, I have compared these with the intensity of gravitational force on Earth's surface, which I arbitrarily set equal to 1.

But it is easy to measure the actual value of the gravitational force at Earth's surface. Since forces are measured by the accelerations they induce, it is only necessary to measure the acceleration of a body dropping, let us say, from the top of the Empire State Building to the ground under the influence of gravity. It turns out that this acceleration and, therefore, the value of the gravitational force (at the equator, at sea level, and corrected for the effects of air resistence) is 980.665 centimeters per second per second, or, in more familiar units, 31.6 feet per second per second.

This means that if an office safe is raised to a height of 5000 feet above the Earth's surface and released, it would fall at the rate of 31.6 feet/sec after one second, twice that (63.2 feet/sec) after two seconds, three times that (94.8 feet/sec) after three seconds, and so on, its rate of fall increasing smoothly with time. (Here and elsewhere in this chapter, I am ignoring the effects of air resistance, which is a subversive influence and a nuisance.)

† This article appeared in December 1958, years before any man was placed in orbit. At the time I prepared the article for inclusion in *Fact and Fancy*, Titov had just orbited seventeen times. In the fourteen years since, many men (and one woman) have been in free fall orbits, sometimes for as long as a quarter of a year.

The equation relating the distance (s) through which a body falls during a time (t) under gravitational acceleration (g) is:

$$s = \frac{1}{2}gt^2 \qquad \text{(Equation 6)}$$

The value of g is, of course, 31.6, and if a body is falling from 5000 feet above Earth's surface, s is 5000. By substituting these figures into Equation 6, it can be solved for t. It turns out that it will take our office safe 17.8 seconds of fall before it splashes into Earth's surface. At the time of contact, it will be moving 17.8×31.6 or 562.5 feet/sec (or 0.106 miles/sec).

(It does not, by the way, matter whether we use a golf ball or an office safe as the falling object. The inertia of an object varies directly with its mass, which means it takes twice the force to accelerate a two-pound weight at a certain rate as it does to accelerate a one-pound weight. But gravitational force also varies with the mass of the falling object. A two-pound weight is attracted to Earth with twice the force of a one-pound weight. Generalizing this, you can see that the end result is that all objects, whatever their mass, experience the same acceleration in a given gravitational field. The effect of air resistance on light objects, such as feathers and leaves, obscures this fact and misled Aristotle—who thought a two-pound weight fell with twice the acceleration of a one-pound weight—and all who followed him down to the time of Galileo.)

The figures on fall under gravity are true in reverse also. If a cannonball is shot directly upward against Earth's gravity, at a velocity of 0.106 miles/sec as it leaves the cannon's mouth, it will travel upward (slowing constantly) for 17.8 seconds and reach a height of 5000 feet before coming to a halt and beginning to fall back.

If our original office safe were raised to a height of 20,000 feet instead of 5000, the time of fall would then be 35.6 seconds and the final velocity is 0.212 miles/sec. And if the cannonball were shot upward at an original velocity of 0.212 miles/sec—but you can see that without my telling you.

It follows, generally, from Equation 6, that the time of fall and the final velocity of a falling object, vary as the square root of the distance of fall, assuming a given constant value of g. It would seem then that the final velocity at contact of office safe and Earth could be as high as you care to make it—by setting the safe to falling from a greater and greater height above the surface.

But there's a catch. I said we must assume "a given constant value of *g*," and that is exactly what we can't do.

The value of *g* varies with distance from the Earth's center, as I explained earlier. In lifting an office safe, or a golf ball, 5000 or even 20,000 feet above Earth's surface, the distance from Earth's center is not significantly changed, and you can work your calculations as though *g* were constant.

But suppose you were to release your object 3950 miles above the surface of the Earth. Up there, the value of *g* is only 0.25 what it is on the surface and the acceleration imposed upon a falling body is likewise only 0.25 what it is here on the surface.

To be sure, the value of *g* increases as the object drops and is a full 1 *g* by the time it is at the collision point. Nevertheless, it takes longer for the object to complete its drop than it would have if the value of *g* were 1 all the way down, and it doesn't hit at as high a velocity as it would if the value of *g* were 1 all the way down.

Every additional thousand miles upward from Earth's surface adds less and less to the final velocity. The result is a converging series where an infinite number of smaller and smaller terms add up to a finite sum. This finite sum, in the case of objects falling toward Earth, is 6.98 miles/sec. This means that if an office safe, or anything else, were to fall from any distance, however great, its final velocity as it struck Earth would never exceed 6.98 miles/sec.

This figure might be called the "maximum final falling velocity," but it isn't. People prefer to look at it in reverse. If a cannonball, a spaceship, or anything else were fired directly upward at a velocity of 6.98 miles/sec (or more), it would continue moving outward indefinitely, if there were no interference from extraneous gravitational fields. (Since a fall even from an infinite distance could not create a final speed of more than 6.98 miles/sec, then the reverse follows: An initial speed of 6.98 miles/sec or more could never be reduced to zero by Earth's gravitation, even if the object traveled forever.)

An object hurled out in this fashion would never return to Earth. It will not have escaped from the influence of the

NASA

MEN ON THE MOON

Fantasies have placed men on the Moon as long ago as Roman times. The first attempt to do so plausibly, using technological expertise, was carried through by Jules Verne in his book *From Earth to the Moon* in the 1860s.

Verne took into account the effect of gravity, of weightlessness, of the airlessness of space. He did his best to describe exactly what the strange environment of space would seem like. He made, of course, the rather bad error of having his astronauts fired from a large gun without noting that the acceleration involved would have killed every man on board at once. Against that, he had the first flight carried through by Americans, he had the spaceship launched from Florida, and he had it made of aluminum.

Once magazine science fiction became part of the American scene, spaceflight was the background of innumerable stories. Every science fiction writer tried his hand at it. As for myself, my very first published story dealt with a spaceship wrecked in the asteroid belt, and in 1939, I published a tale of the first flight to the Moon —which I placed in 1978.

Finally, in 1969 (a decade after this essay first appeared) science fiction writers and readers had the odd experience of watching astronauts land on the Moon and seeing something happen exactly as they had always imagined it would. The picture here is that of an Apollo 17 astronaut walking the Moon in December 1972 but there is absolutely nothing in the photograph to distinguish it from imaginative illustrations of stories written in December 1932.

One thing that no science fiction writer visualized, however, as far as I know, was that the landings on the Moon would be watched by people on Earth by way of television.

Earth's gravitational field (which will be slowing it constantly) but it will have escaped Earth itself.

So the velocity of 6.98 miles a second is the "escape velocity" for Earth.

The value of the escape velocity varies with the mass of the attracting body and the distance from its center as follows:

$$v = 6.98\sqrt{m/d} \qquad \text{(Equation 7)}$$

where v is the escape velocity, m is the mass of the attracting body in "Earth-masses" and d the distance to the center of the attracting body in "Earth-radii." The factor 6.98 allows the escape velocity to come out in miles per second.

The Moon, for instance, has a mass equal to 0.0123 "Earth-masses" and, at its surface, the distance from its center is 0.273 "Earth-radii." The escape velocity from the Moon's surface is therefore $6.98 \times \sqrt{0.0123/0.273}$, or 1.49 miles/sec.

The escape velocities at the surface of any body in the

Solar System can be similarly calculated and the results are presented in Table 3.

One caution: Escape velocity is required for escape from a planet only where unpowered (i.e. "ballistic") flight is concerned. If you are in a spaceship under constant power, you can move any finite distance from Earth at any velocity below escape velocity but above zero, provided you have fuel enough. (In the same way, you cannot jump to a second story window at a bound unless the initial thrust of your leg muscles against the ground is great enough—which is more than you can manage—but you can nevertheless walk up two flights of stairs as slowly as you please.)

TABLE 3 Escape Velocities at the Surface of Bodies in the Solar System

Astronomical Body	*Mass* (EARTH-MASSES)	*Radius* (EARTH-RADII)	*Escape Velocity* (MILES PER SECOND)
Jupiter (pole)	318.	10.5	38.4
Jupiter (equator)	318.	11.2	37.3
Saturn (pole)	95.2	8.5	23.4
Saturn (equator)	95.2	9.5	22.1
Neptune	17.3	3.4	15.8
Uranus	14.5	3.7	13.9
Earth	1.0	1.0	6.98
Venus	0.82	0.96	6.46
Mars	0.11	0.525	3.20
Mercury	0.054	0.380	2.64
Ganymede	0.026	0.395	1.80
Moon	0.0123	0.273	1.49

And yet escape from Earth may be not entirely escape, either. I said earlier that an object hurled from Earth at more than escape velocity would move outward forever "if there were no interference from extraneous gravitational fields."

But, of course, there *is* such interference. Consider the Sun, for instance, which so far we haven't done.

The Sun has a mass that is equal to 330,000 "Earth-masses" and a radius equal to 109 "Earth-radii." Using Equation 7, the escape velocity from the Sun's surface turns out to be a tidy 385 miles/sec.

From Earth, however, the distance to the Sun's center is about 23,000 "Earth-radii." Substituting that figure for d in Equation 7, and leaving m at 330,000 "Earth-masses," it turns out that the escape velocity from the Sun at Earth's distance is 26.4 miles/sec.

This is four times as high as the escape velocity from Earth itself. In other words, a missile shot out from Earth and attaining a velocity of 6.98 miles/sec by the time the rocket thrust is expended, may be free of the Earth, but *it is not free of the Sun*. It will not recede forever after all, but will take up an orbit about the Sun.

To escape from the Solar System altogether, a speed of 26.4 miles/sec must be attained in ballistic flight. To be sure, in powered flight, we don't have to attain escape velocity; we can just keep the engines going. However, the escape velocity is a measure of the amount of energy we must use to break the gravitational chains in any fashion. So you see, it is the Solar prison bars that block our way to the stars far more than Earth's puny fence.

The only consolation is that, for the moment, the Moon and planets are enough of a challenge. The stars can wait.††

†† They only waited fourteen years. On March 2, 1972, the probe "Pioneer 10" was launched toward Jupiter. Its speed at launching was nine miles per second, not enough to free it of the Sun's grip. When it reached Jupiter in December, however, it whipped about that planet and Jupiter's great gravitational field accelerated it to a speed greater than the Sun's escape velocity at the distance of Jupiter. Pioneer 10 is the first man-made object which will (barring accidental collision with some asteroid or the like) escape from the Solar System. In 1984 it will pass the orbit of Pluto and will then wander on through interstellar space, as far as we can tell, forever.

Four
OF CAPTURE AND ESCAPE

Since January 2, 1959, the Soviet Union and the United States have sent up a number of missiles which were notable for three things:

(1) They reached and passed the orbit of the Moon.

(2) They were not captured by the Moon; that is, they did not take up a closed orbit about the Moon alone.*

(3) They took up a closed orbit about the Sun and became artificial planets.†

I'd like to consider each of these points in turn.

First, what does it take to reach the orbit of the Moon by means of a ballistic missile? (A ballistic missile is any projectile which receives an initial impulse of some sort and thereafter moves under the influence of gravitational forces only.)

If such a missile is fired straight up (i.e. directly away from Earth's center) the maximum height it will reach will depend (a) on the strength of the initial impulse upward and (b) the strength of Earth's gravitational pull downward.

Naturally, the greater the initial impulse upward, the greater the height reached. You might expect that doubling the initial impulse will double the height reached, but that is too pessimistic. It would be so if the gravitational force remained constant all the way up, but it does not. The higher the missile reaches, the weaker the gravitational drag upon it. The second half of its climb meets less resistance therefore and is correspondingly extended.

* In 1966, seven years after this article appeared, the first of the "Lunar Orbiter" satellites was launched. They *were* placed in orbit about the Moon, as satellites of a satellite. In 1971 "Mariner 9" was placed in orbit around Mars, the first man-made satellite of another planet.

† Pioneer 10 was the first to escape the grip of the Sun (see footnote on the previous page).

Consequently, doubling the initial impulse *more* than doubles the maximum height reached, and the more you increase the initial impulse, the more drastically do you increase the maximum height reached.††

Table 4 gives the maximum height attained for various initial velocities of the missile. The initial velocity is a measure of the strength of the push given the missile. (Naturally, there are complicating factors. There is air resistance; there is the fact that the push of the rocket motors isn't administered instantaneously, but is spread over several minutes, and so on. Since we're all friends here, I'm taking the privilege of ignoring such matters and leaving them to the missile engineers, who are most welcome to them.)

TABLE 4

Initial Velocity of Missile		*Maximum Height above Earth's Surface*
(MILES PER SECOND)	(MILES PER HOUR)	(MILES)
1	3,600	80
2	7,200	350
3	10,800	900
4	14,400	1,940
5	18,000	4,180
5.5	19,800	6,450
6.0	21,600	11,100
6.5	23,400	25,800
6.6	23,760	34,300
6.7	24,120	46,300
6.8	24,480	73,600
6.85	24,660	102,800
6.90	24,840	170,000
6.92	24,910	221,000
6.95	25,020	454,000
6.98	25,130	∞

Notice how quickly the maximum height increases, especially at speeds higher than 6 miles a second, or, if you prefer, 21,600 miles an hour. (I have always had a liking for the use of "miles per second" as the unit for high

†† I know this duplicates what I said in the previous chapter, but this chapter appeared, originally, five months after the previous chapter did.

velocities, but to a nation of automobile drivers "miles per hour" seems more natural. Besides, newspapers and allied information-mongers use "miles per hour" exclusively, perhaps because larger and flashier numbers are involved. So I'll use both units throughout. I just wish to warn you, though, that 21,600 miles an hour may sound flashier than 6 miles a second, but the two are entirely equivalent.)

A missile leaving Earth with an initial velocity of 6.92 miles a second (24,912 miles an hour) will reach a height of 220,000 miles before coming to a halt and beginning to fall back. This is just about the distance of the Moon at its closest approach ("perigee") to the Earth.

If, however, the missile leaves Earth at a velocity of 6.90 miles a second (24,840 miles an hour), it falls 50,000 miles short of the Moon. A difference of 0.02 miles a second (72 miles an hour) to begin with means a 50,000 mile discrepancy to end with.

It is for this reason that when one of our early Moon-probes only reached a third of the way to the Moon, it did *not* mean we had only attained a third of the necessary velocity. Actually, we had attained over 98 per cent of the necessary velocity. It's just that the last per cent or so is what carries the missile the remaining two thirds of the way to the Moon.

To go back to Table 4, a missile leaving Earth at a velocity of 6.98 miles a second (25,130 miles an hour—or something like 216 miles an hour faster than is required to reach the Moon's orbit) has no maximum height. If you like, its maximum height is infinite, symbolized as ∞ in the table. Such a missile would move away from Earth forever, assuming there is no interference from gravitational fields of other bodies. The velocity of 6.98 miles a second (25,130 miles an hour) is therefore the "escape velocity" from Earth's surface.

Imagine a missile that has left the Earth's surface at just the escape velocity. As it travels away from the Earth, its velocity decreases inversely as the square root of its distance from Earth's center. (When the distance has been multiplied by 4, the velocity has been decreased by 2.) The result is shown in Table 5.

Earth's gravitational pull is constantly decreasing the missile's velocity, but with increasing distance, the pull loses power and decreases the velocity at a slower and slower rate. The velocity therefore gets closer and closer to zero as the missile recedes from Earth, but never quite gets to zero.

If the missile had left at less than the escape velocity, Earth's gravity would have managed to bring the missile's velocity to zero at some finite distance and the missile would then fall back. If the missile leaves at a speed greater than the escape velocity, its velocity decreases and decreases with distance but never falls below a certain velocity, greater than zero, however far it travels. (All this assumes the presence of no other gravitational fields in the Universe, gumming up the works.)

Let's put it another way. A missile leaving Earth at a velocity less than the escape velocity follows an elliptical orbit. An ellipse is a closed curve, so that the missile does not depart more than a certain distance from the Earth. If the elliptical orbit happens to intersect Earth's surface, the missile crashes its first time round, as our first Moon-probes did. If the elliptical orbit does not intersect the Earth's surface, artificial satellites are the result.

A missile leaving Earth at a velocity just equal to escape velocity takes up a parabolic orbit. A parabola is an open curve that never turns back on itself. Consequently, any object leaving Earth on a parabolic orbit never returns, barring the interference of the gravitational fields of other heavenly bodies.

If a missile leaves Earth at more than escape velocity, it follows a hyperbolic orbit. A hyperbola is also an open curve—even more open than a parabola, in a manner of speaking—so again the missile never returns.

Returning now to Table 5 (this gets complicated but I'm slowly building up a line of argument which, I hope, I can put to good use) I want to point out a special significance of the "velocity" column. The velocity of the missile which began at escape velocity remains at escape velocity throughout!

To be sure, the actual velocity of the missile is continually decreasing as its distance from Earth increases; but so does the escape velocity. And the escape velocity keeps pace throughout; for it, too, varies inversely as the square root of the distance from Earth.

Suppose you were to start from scratch at a distance 8000 miles from Earth's center, which is just about 4000

TABLE 5

Distance from the Center of the Earth (MILES)	*Velocity of Missile Fired at Escape Velocity* (MILES PER SECOND)	(MILES PER HOUR)
4,000 (Earth's surface)	6.98	25,130
8,000	4.93	17,800
12,000	4.04	14,500
16,000	3.49	12,500
20,000	3.12	11,210
40,000	2.21	7,950
80,000	1.56	5,620
120,000	1.27	4,570
160,000	1.10	3,960
221,000 (Moon at perigee)	0.95	3,410
253,000 (Moon at apogee)	0.88	3,160
400,000	0.70	2,510
1,000,000	0.44	1,580
∞	0.00	0

miles above Earth's surface. (Imagine, in other words, that you were on top of a mountain—a mythical one—4000 miles high.) Up there Earth's gravitational pull would be only one fourth what it is at sea level. There would be that much less drag on the missile and a smaller initial velocity would suffice to kick it into a parabolic orbit. To be exact, 4.93 miles a second (17,800 miles an hour) would suffice.

And from a mountain 80,000 miles high, an initial velocity of 1.56 miles a second (5620 miles an hour) would suffice. And from a mountain 1,000,000 miles high, 0.44 miles a second (1580 miles an hour) would suffice.

But at no finite distance from Earth, however great, would escape velocity actually be zero. At any finite distance, an object completely at rest with respect to Earth, would start moving toward the Earth in response to its gravitational pull—provided no other gravitational field interferes. To prevent the object from falling to Earth, some definite opposing push is needed; perhaps an infinitesimally small one if the distance is great, but some push is needed.

All this holds true for a missile (or a meteor) passing close by Earth from some outer-space starting point.

Suppose a meteor passed Earth at a distance of 120,000 miles from its center and had a velocity (with respect to Earth) of less than 1.27 miles a second (4570 miles an hour). Since the meteor's velocity is less than the escape velocity at its point of approach, it is forced into an elliptical orbit about the Earth. It is captured.

If its velocity were exactly 1.27 miles a second (4570 miles an hour) it would take up a parabolic orbit; if its velocity were greater it would take up a hyperbolic orbit. In both these latter cases, its direction of travel would be changed and it would curve about Earth more or less sharply. But in neither case would it be captured. It would go shooting off into space never to return.

Of course, both parabolic and hyperbolic orbits travel about the *center* of the Earth as a focus. If the meteor is aimed in such a fashion that its new orbit will pass within 4000 miles of the Earth's center, it will intersect Earth's surface. The meteor will then enter our atmosphere and flame to death. However, hitting the Earth is not the same as being captured by the Earth.

Since escape velocity increases with decreasing distance from Earth, a meteor is more likely to be captured if it passes close to the Earth, than if it passes at a distance. A meteor traveling at a velocity of 3.12 miles a second (11,210 miles per hour) relative to the Earth, will be captured if it passes Earth at a distance of less than 20,000 miles, but not if it passes Earth at a distance greater than that. Below 20,000 miles its velocity is less than escape velocity; above, it is higher than escape velocity.

The more massive a planet is, the higher its escape velocity at all distances, and the more likely it is to capture invading meteors and planetoids. Jupiter, for instance, with a mass 318 times that of Earth has an escape velocity at its surface of 37.3 miles per second (134,000 miles an hour). Since Jupiter's surface is some 40,000 miles from its center, the comparable escape velocity in the case of Earth is only 2.21 miles a second (7950 miles an hour). At a distance of 1,000,000 miles from Jupiter's center, the escape velocity is 13.2 miles a second (47,500 miles an hour) as compared to 0.44 miles a second (1580 miles an hour) for a comparable distance from Earth.

It is not surprising then that the seven* outermost of Jupiter's twelve satellites are generally considered to be captured planetoids. But if a more massive planet is a more efficient capturer of wandering objects, a less massive astronomical body should be a less efficient capturer. That

* Eight, as of 1974.

The Granger Collection

JUPITER

Jupiter was named for the king of the gods. The ancient Babylonians named it Marduk, after the ruler of their pantheon. The reason is plain, for it was at once bright and free. Of the planets, only Venus is brighter, but it always remains close to the Sun and is never present in the midnight sky. Such weakness and dependency befits a goddess—the most beautiful one, of course.

Jupiter, however, nearly as bright as Venus, can appear

at any distance from the Sun and seems truly independent. On a moonless midnight, Jupiter, if present, is the brightest object in the sky. Naturally, it should be named for the king of the gods. The Greeks followed the Babylonian lead and called the planet Zeus, while the Romans (and we) call it Jupiter.

It is an odd coincidence that once the telescope was invented and Jupiter could be studied in detail, it turned out that the planet named for the king of the gods was indeed the largest, by far, in the Solar System; that it had eleven times the diameter of the Earth, over three hundred times the mass of our planet, and was well over a thousand times as voluminous.

Well, it was not entirely coincidence. The size and the brightness go together. At least part of the reason Jupiter is so bright is that it is so large.

In this photograph, taken by the 200-inch telescope, you can see the belts and the Great Red Spot. Just a little beyond the planetary globe is the satellite Ganymede, and in the planetary polar region you can see the shadow it casts.

Far more spectacular photographs have been taken recently by the Jupiter probes, Pioneer 10 and 11, and we now think of the planet as a huge drop of extremely hot and somewhat impure liquid hydrogen.

brings us to the Moon, which is only $\frac{1}{81}$ as massive as the Earth and should therefore be a very poor capturer of meteors and assorted debris such as missiles.

The escape velocity from the Moon's surface is a mere 1.49 miles a second (5360 miles an hour) and this falls off, in the usual way, in inverse ratio to the square root of the distance, from the Moon's center. The escape velocity at various distances from the Moon is given in Table 6.

To be captured by the Moon, a missile must pass the Moon at a velocity less than the escape velocity at that distance. What's more, the velocity involved is the velocity relative to the Moon, not relative to the Earth.

The Moon, you see, is itself moving at a velocity of about 0.64 miles a second (2300 miles an hour) with respect to the Earth. Suppose, then, a missile shot from Earth at 6.92 miles a second (24,912 miles an hour) just makes it to the Moon's orbit and hangs momentarily suspended at zero velocity (with respect to the Earth) at a distance of 4500 miles from the Moon's surface (5500 from its center).

The Moon, however, is retreating from it, or advancing toward it, or passing to one side of it (depending on the exact position of the missile with respect to the Moon) at 0.64 miles a second (2300 miles an hour), so that is the missile's velocity *relative to the Moon.* This velocity is just a bit over the Moon's escape velocity at the distance of 5500 miles from its center.

If the missile had been fired with a greater initial velocity, so that it was still moving at some velocity or other when it reached the Moon's orbit, its velocity relative to the Moon would be greater still.

It follows then that any missile that misses the Moon's center by 5500 miles or more cannot be captured by the Moon and will not move into an orbit about the Moon, no matter how slowly the missile is going. The respective motions may be such that the missile may *hit* the Moon, as did the Soviet Union's Lunik II, but that's another thing. It may hit the Moon but it won't be captured by the Moon in the sense that it will go into a closed orbit about it.

TABLE 6

Distance from the Center of the Moon (MILES)	*Velocity of Missile Fired from Moon at Escape Velocity* (MILES PER SECOND)	(MILES PER HOUR)
1,000 (Moon's Surface)	1.49	5,360
1,500	1.21	4,360
2,000	1.06	3,820
2,500	0.94	3,380
3,000	0.86	3,100
3,500	0.80	2,880
4,000	0.74	2,560
4,500	0.70	2,520
5,000	0.66	2,375
5,500	0.63	2,270
∞	0.00	0

A missile fired from Earth at escape velocity will pass the Moon (at perigee) at 0.95 miles a second (3410 miles an hour). Thanks to the Moon's own motion, which will be roughly at right angles to that of the missile, the missile's velocity with respect to the Moon will be 1.15 miles a second (4140 miles an hour). This is the Moon's escape velocity at a distance of about 1600 miles from its center. Such a missile would therefore have to come within 600 miles of the Moon's surface before it can be captured and go into an orbit about the Moon.

A missile fired from Earth at 7.37 miles per second (26,500 miles an hour) will pass the Moon at a velocity of 1.34 miles per second (4820 miles an hour) with respect to the Earth, but a velocity of 1.49 miles per second (5360 miles an hour) with respect to the Moon. This is the Moon's escape velocity at its surface. A missile fired from Earth at this velocity or above cannot be captured by the Moon, no matter how close to the Moon it passes, not even if it grazes its surface. (I repeat, it can *hit* the Moon, but again I repeat, that's a different thing.)

So the limits for success are narrow indeed. A missile must be fired at a velocity of at least 6.92 miles per second (24,910 miles an hour) or it won't reach the Moon; and it must be fired at a velocity of less than 7.37 miles per second (26,500 miles an hour) or it can't be captured by the Moon. And even within that narrow range of velocities, capture by the Moon is only possible if the missile passes quite close to the Moon. A miss of not more than 4500 miles from the Moon's surface is the maximum, and this leeway rapidly decreases as you approach the upper limit of the permissible range.

In fact, ballistic missiles are so hard to place into an orbit about the Moon that I wonder if it's even sensible to try. It might be better to make the missile non-ballistic. That is, to supply a final delayed rocket blast which could be set off by radio at such a time and in such a direction as to decrease the velocity of the missile relative to the Moon and make it capturable.†

This brings us to the final point I raised at the beginning of the article, the question of orbiting about the Sun.

As I pointed out in the chapter "Catching Up with Newton," the escape velocity from the Sun, even way out here at Earth's orbit, 93 million miles from the Sun, is still 26.4

† And, of course, that was exactly what was done in placing Lunar Orbiters in orbit about the Moon and, for that matter, in placing manned vessels in orbit about the Moon prior to landing.

The Granger Collection

THE MOON

What an amazing satellite we have! By the purest coincidence, its distance and size are such as to make its apparent size almost identical with that of the Sun. Every once in a while, therefore, the Moon can get in front of the Sun to offer us an unbelievably spectacular celestial show.

The Earth has no business possessing such a Moon. It is too huge—over a quarter Earth's diameter and about ⅟81 its mass. No other planet in the Solar System has even nearly so large a satellite.

The Moon's size makes it clearly visible in the sky as a circle rather than as a dot of light. What's more, as it changes position relative to the Sun, it goes through a succession of phases endlessly repeated. The passing phases can be used as a calendar and the relationship with the phases is complex enough to make it urgent for mankind to develop mathematics, astronomy, even religion.

More than that, the Moon is close enough and large enough to reveal blotches, particularly at full Moon, as here indicated. It is the only heavenly object to possess features visible to the unaided eye. It made the Moon seem imperfect and the phases made it clear that it shone only by reflected light. The Moon was a *world* and from early times it seemed that it, like the Earth, might be inhabited—even if only by a legendary "man in the Moon."

When Galileo first used the telescope in 1609, the Moon was shown to possess mountains, plains, and craters. It was a world and it lured human beings on to dream of spaceflight. What a wonderful primer it was—only 400,000 kilometers away and reachable in three days with even the clumsiest of chemical space-drives!

miles per second (95,040 miles an hour). I left it at that point then, but let's carry it further now.

The figure 26.4 miles a second (95,040 miles an hour) refers, of course, to velocity relative to the Sun. If the Earth were at rest with respect to the Sun, we would have to fire a missile at that initial velocity to free it of the Sun's grip. However, the Earth is *not* at rest with respect to the Sun, but travels in an orbit about the Sun at a velocity of 18.5 miles a second (66,600 miles an hour).

Suppose, then, we were to fire a missile in the direction of the Earth's motion. It would already be traveling 18.5 miles a second (66,600 miles an hour) with respect to the Sun before it started. Giving it additional velocity would raise the figure (like flying an airplane downwind). A velocity, relative to the Earth, of 7.9 miles a second (28,440 miles an hour) would just suffice to raise the missile's velocity to the point where it could escape the Solar System altogether, provided it didn't hit something on the way.

This is the most economical way of freeing a missile from the grip of both Earth and Sun.

If a missile were fired at right angles to Earth's motion, either directly toward or away from the Sun, it would receive some but not all the benefit of Earth's motion (like an airplane flying cross-wind). The missile would have to be fired at an initial velocity of 18.8 miles a second (67,680 miles an hour) to attain to Solar System escape.

If it were fired in the direction opposite to the Earth's motion, Earth's motion would then not be helping but hindering. The missile would require the full initial velocity of escape from the Sun plus enough more to neutralize the Earth's motion (like an airplane flying upwind). For a missile so fired to escape would require an initial velocity of 44.9 miles a second (161,600 miles an hour).

The first successful Moon-probe was fired at a time when the Moon was in "last quarter." At this time, the Moon is directly ahead of the Earth in their path around the Sun, so the probe was fired in the direction of Earth's motion. Nevertheless, if we remember that the probable initial velocity of the missile might have been as high as 7.5 miles a second (27,000 miles an hour) this is still insufficient to allow escape from the Sun, and the missile remained in orbit about the Sun.

To be sure, it has a higher velocity than the Earth has so that its orbit bellies out into the space between Earth and Mars. (Since the missile's velocity is higher than Earth's, it makes a slightly more effective attempt, so to speak, to get away from the Sun, and it gets halfway to Mars before the Sun pulls it back.) As a result, the missile's year is 15 months long, rather than 12 months long as is our Earth's.

The two orbits cross, however, and it is conceivable that someday both missile and Earth may be at the crossing point simultaneously, in which case the missile will finally come home.

One last question. Was there any chance that a missile such as Lunik I or Pioneer IV might have fallen into the Sun?

Well, let's see what's required to hit the Sun. Suppose you aimed a missile directly at the Sun. It would travel toward the Sun, yes, but at the same time it would retain Earth's motion of 18.5 miles a second (66,600 miles an hour) in a direction at right angles to its aimed line of motion at the Sun. Its overall motion would be a combination of both component motions. Earth's sidewise motion would therefore carry the missile around the Sun in an elliptical orbit, if its initial velocity with respect to Earth were less than 18.8 miles a second (67,680 miles an hour)—this being the Solar escape velocity for a missile fired at right angles to Earth's motion.

If the missile were fired at exactly the escape velocity, the component due to Earth's motion would carry the missile about the Sun in a parabolic orbit; if it were greater than escape velocity it would go about the Sun in a hyperbolic orbit.

The greater the velocity in the direction of the Sun, the flatter the hyperbola and the closer it would approach the center of the Sun at its closest approach. If you aimed at the center of the Sun, no velocity short of the infinite would enable you to hit the center, thanks to the sidewise component of motion.

Of course, why aim at the Sun's center? Why not aim to one side of it, allowing Earth's motion to bring the missile to the Sun; instead of aiming at it and allowing Earth's motion to carry the missile past it. (This is like allowing for the wind when you aim a gun.)

The most economical way to neutralize Earth's motion is to shoot the missile in a direction directly opposite to that motion. If the missile is then fired at a velocity of just 18.5 miles a second (66,600 miles an hour), Earth's motion with respect to the Sun is neutralized. The missile is, in fact, then at rest with respect to the Sun, and it will proceed to fall into the Sun under the inexorable pull of Solar gravity.

If a missile is fired in this opposite direction to Earth's motion (that is, at the Moon at "first quarter") at less than this speed, its motion with respect to the Sun is still less than that of the Earth. It would not fall into the Sun, but it would approach it more closely than does Earth, its orbit moving in toward that of Venus. We would then have a Venus-probe as in the case of Pioneer V.

There's a lesson here. A spaceflight to Mars must start off in the direction of Earth's motion, while one to Venus must start off in the direction opposite Earth's motion—at least if we want to make economical use of the motion we are already blessed with from birth.

Five
FIRST AND REARMOST

When I was in junior high I used to amuse myself by swinging on the rings in gym. (I was lighter then, and more foolhardy.) On one occasion I grew weary of the exercise, so at the end of one swing I let go.

It was my feeling at the time, as I distinctly remember, that I would continue my semicircular path and go swooping upward until gravity took hold; and that I would then come down light as gossamer, landing on my toes after a perfect *entrechat*.

That is not the way it happened. My path followed nearly a straight line, tangent to the semicircle of swing at the point at which I let go. I landed good and hard on one side.

After my head cleared, I stood up* and to this day that is the hardest fall I have ever taken.†

I might have drawn a great deal of intellectual good out of this incident. I might have pondered on the effects of inertia; puzzled out methods of summing vectors; or deduced some facts about differential calculus.

However, I will be frank with you. What really impressed itself upon me was the fact that the force of gravity was both mighty and dangerous and that if you weren't watching every minute, it would clobber you.

Presumably, I had learned that, somewhat less drastically, early in life; and presumably, every human being who ever got onto his hind legs at the age of a year or less and promptly toppled, learned the same fact.

In fact, I have been told that infants have an instinctive

* People react oddly. After I stood up, I completely ignored my badly sprained (and possibly broken, though it later turned out not to be) right wrist and lifted my untouched left wrist to my ear. What worried me was whether my wristwatch was still running.

† I took a harder fall (slipping on ice) four years after this article first appeared. I didn't break anything that time either, I'm relieved to say.

fear of falling, and that this arose out of the survival value of having such an instinctive fear during the tree-living aeons of our simian ancestry.

We can say, then, that gravitational force is the first force with which each individual human being comes in contact. Nor can we ever manage to forget its existence, since it must be battled at every step, breath, and heartbeat. Never for one moment must we cease exerting a counterforce.

It is also comforting that this mighty and overwhelming force protects us at all times. It holds us to our planet and doesn't allow us to shoot off into space. It holds our air and water to the planet too, for our perpetual use. And it holds the Earth itself firmly in its orbit about the Sun, so that we always get the light and warmth we need.

What with all this, it generally comes as a rather surprising shock to many people to learn that gravitation is *not* the strongest force in the universe. Suppose, for instance, we compare it with the electromagnetic force that allows a magnet to attract iron or a proton to attract an electron. (The electromagnetic force also exhibits repulsion, which gravitational force does not, but that is a detail that need not distress us at this moment.)

How can we go about comparing the relative strengths of the electromagnetic force and the gravitational force?

Let's begin by considering two objects alone in the universe. The gravitational force between them, as was discovered by Newton, can be expressed by the following equation:

$$F_g = \frac{Gmm'}{d^2} \qquad \text{(Equation 8)}$$

where F_g is the gravitational force between the objects; m is the mass of one object; m' the mass of the other; d the distance between them; and G a universal "gravitational constant."

We must be careful about our units of measurement. If we measure mass in grams, distance in centimeters, and G in somewhat more complicated units, we will end up by determining the gravitational force in something called "dynes." (Before I'm through with this chapter, the dynes will cancel out, so we need not, for present purposes, consider the dyne anything more than a one-syllable noise.)

Now let's get to work. The value of G is fixed (as far as

we know) everywhere in the universe.†† Its value in the units I am using is 6.67×10^{-8}. If you prefer long zero-riddled decimals to exponential figures, you can express *G* as 0.0000000667.*

Let's suppose, next, that we are considering two objects of identical mass. This means that $m = m'$, so that mm' becomes mm, or m^2. Furthermore, let's suppose the particles to be exactly 1 centimeter apart, center to center. In that case $d = 1$, and $d^2 = 1$ also. Therefore, Equation 8 simplifies to the following:

$$F_g = 0.0000000667\, m^2 \qquad \text{(Equation 9)}$$

We now proceed to the electromagnetic force, which we can symbolize as F_e.

Exactly one hundred years after Newton worked out the equation for gravitational forces, the French physicist Charles Augustin de Coulomb was able to show that a very similar equation could be used to determine the electromagnetic force between two electrically charged objects.

Let us suppose, then, that the two objects for which we have been trying to calculate gravitational forces also carry electric charges, so that they also experience an electromagnetic force. In order to make sure that the electromagnetic force is an attracting one and is therefore directly comparable to the gravitational force, let us suppose that one object carries a positive electric charge and the other a negative one. (The principle would remain even if we used like electric charges and measured the force of electromagnetic repulsion, but why introduce distractions?)

According to Coulomb, the electromagnetic force between the two objects would be expressed by the following equation:

$$F_e = \frac{qq'}{d^2} \qquad \text{(Equation 10)}$$

where *q* is the charge on one object, *q'* on the other, and *d* is the distance between them.

†† There is more and more speculation to the effect that the value of G may be very slowly decreasing as the Universe expands. True or not, that does not affect the argument as it relates to the present.

* In 1974, the value was measured as 0.000000066720.

If we let distance be measured in centimeters and electric charge in units called "electrostatic units" (usually abbreviated "esu"), it is not necessary to insert a term analogous to the gravitational constant, provided the objects are separated by a vacuum. And, of course, since I started by assuming the objects were alone in the universe, there is necessarily a vacuum between them.

Furthermore, if we use the units just mentioned, the value of the electromagnetic force will come out in dynes.

But let's simplify matters by supposing that the positive electric charge on one object is exactly equal to the negative charge on the other, so that $q = q'$† which means that $qq' = qq = q^2$. Again, we can allow the objects to be separated by just one centimeter, center to center, so that $d^2 = 1$. Consequently, Equation 10 becomes:

$$F_e = q^2 \qquad \text{(Equation 11)}$$

Let's summarize. We have two objects separated by one centimeter, center to center, each object possessing identical charge (positive in one case and negative in the other) and identical mass (no qualifications). There is both a gravitational and an electromagnetic attraction between them.

The next problem is to determine how much stronger the electromagnetic force is than the gravitational force (or how much weaker, if that is how it turns out). To do this we must determine the ratio of the forces by dividing (let us say) Equation 11 by Equation 9. The result is:

$$\frac{F_e}{F_g} = \frac{q^2}{0.0000000667\ m^2} \qquad \text{(Equation 12)}$$

A decimal is an inconvenient thing to have in a denominator, but we can move it up into the numerator by taking its reciprocal (that is, by dividing it into 1). Since 1 divided by 0.0000000667 is equal to 1.5×10^7, or 15,000,-000, we can rewrite Equation 12 as:

† We could make one of them negative to allow for the fact that one object carries a negative electric charge. Then we could say that a negative value for the electromagnetic force implies an attraction and a positive value a repulsion. However, for our purposes, none of this folderol is needed. Since electromagnetic attraction and repulsion are but opposite manifestations of the same phenomenon, we shall ignore signs.

$$\frac{F_e}{F_g} = \frac{15{,}000{,}000\ q^2}{m^2} \qquad \text{(Equation 13)}$$

or, still more simply, as:

$$\frac{F_e}{F_g} = 15{,}000{,}000(q/m)^2 \qquad \text{(Equation 14)}$$

Since both F_e and F_g are measured in dynes, then in taking the ratios we find we are dividing dynes by dynes. The units, therefore, cancel out, and we are left with a "pure number." We are going to find, in other words, that one force is stronger than the other by a fixed amount; an amount that will be the same whatever units we use or whatever units an intelligent entity on the fifth planet of the star Fomalhaut wants to use. We will have, therefore, a universal constant.

In order to determine the ratios of the two forces, we see from Equation 14 that we must first determine the value of q/m; that is, the charge of an object divided by its mass. Let's consider charge first.

All objects are made up of subatomic particles of a number of varieties. These particles fall into exactly three classes, however, with respect to electric charge:

1) Class A are those particles which, like the neutron and the neutrino, have no charge at all. Their charge is 0.

2) Class B are those particles which, like the proton and the positron, carry a positive electric charge. But all particles which carry a positive electric charge invariably carry the same quantity of positive electric charge whatever their differences in other respects (at least as far as we know). Their charge can therefore be specified as +1.

3) Class C are those particles which, like the electron and the anti-proton, carry a negative electric charge. Again, this charge is always the same in quantity. Their charge is −1.††

You see, then, that an object of any size can have a net electric charge of zero, provided it happens to be made up

†† There are some particles—quarks—that have been postulated and are supposed to have fractional charges of ⅓ or ⅔. Quarks, however, have not, at this time of writing, been detected.

of neutral particles and/or equal numbers of positive and negative particles.

For such an object $q = 0$, and no matter how large its mass, the value of q/m is also zero. For such bodies, Equation 14 tells us, F_e/F_g is zero. The gravitational force is never zero (as long as the objects have any mass at all) and it is, therefore, under these conditions, infinitely stronger than the electromagnetic force and need be the only one considered.

This is just about the case for actual bodies. The over-all net charge of Earth and the Sun is virtually zero, and in plotting the Earth's orbit it is only necessary to consider the gravitational attraction between the two bodies.

Still, the case where $F_e = 0$ and, therefore, $F_e/F_g = 0$ is clearly only one extreme of the situation and not a particularly interesting one. What about the other extreme? Instead of an object with no charge, what about an object with maximum charge?

If we are going to make charge maximum, let's first eliminate neutral particles which add mass without charge. Let's suppose, instead, that we have a piece of matter composed exclusively of charged particles. Naturally it is of no use to include charged particles of both varieties, since then one type of charge would cancel the other and total charge would be less than maximum.

We will want one object then, composed exclusively of positively charged particles and another exclusively of negatively charged particles. We can't possibly do better than that as a general thing.

And yet while all the charged particles have identical charges of either $+1$ or -1, as the case may be, they possess different masses. What we want are charged particles of the smallest possible mass. In that case the largest possible individual charge is hung upon the smallest possible mass, and the ratio q/m is at a maximum.

It so happens that the negatively charged particle of smallest mass is the electron and the positively charged particle of smallest mass is the positron. For those bodies, the ratio q/m is greater than for any other known object (nor have we any reason, as yet, for suspecting that any object of higher q/m remains to be discovered).

Suppose, then, we start with two bodies, one of which contains a certain number of electrons and the other the same number of positrons. There will be a certain electromagnetic force between them and also a certain gravitational force.

If you triple the number of electrons in the first body and triple the number of positrons in the other, the total charge triples for each body and the total electromagnetic force, therefore, becomes 3 times 3, or 9 times greater. However, the total mass also triples for each body and the total gravitational force also becomes 3 times 3, or 9 times greater. While each force increases, they do so to an equal extent, and the ratio of the two remains the same.

In fact the ratio of the two forces remains the same, even if the charge and/or mass on one body is not equal to the charge and/or mass on the other; or if the charge and/or mass of one body is changed by an amount different from the charge in the other.

Since we are concerned only with the ratio of the two forces, the electromagnetic and the gravitational, and since this remains the same, however much the number of electrons in one body and the number of positrons in the other are changed, why bother with any but the simplest possible number—one?

In other words, let's consider a single electron and a single positron separated by exactly 1 centimeter. This system will give us the maximum value for the ratio of electromagnetic force to gravitational force.

It so happens that the electron and the positron have equal masses. That mass, in grams (which are the mass-units we are using in this calculation) is 9.1×10^{-28} or, if you prefer, 0.00000000000000000000000000091.*

The electric charge of the electron is equal to that of the positron (though different in sign). In electrostatic units (the charge-units being used in this calculation), the value is 4.8×10^{-10}, or 0.00000000048.

To get the value q/m for the electron (or the positron) we must divide the charge by the mass. If we divide 4.8×10^{-10} by 9.1×10^{-28}, we get the answer 5.3×10^{17} or 530,000,000,000,000,000.

But, as Equation 14 tells us, we must square the ratio q/m. We multiply 5.3×10^{17} by itself and obtain for $(q/m)^2$ the value of 2.8×10^{35}, or 280,000,000,000,000,-000,000,000,000,000,000,000.

* The best figure currently available is 9.109534×10^{-28}.

Again, consulting Equation 14, we find we must multiply this number by 15,000,000, and then we finally have the ratio we are looking for. Carrying through this multiplication gives us 4.2×10^{42}, or 4,200,000,000,000,000,000,000,000,000,000,000,000,000,000.

We can come to the conclusion, then, that the electromagnetic force is under the most favorable conditions, over four million trillion trillion trillion times as strong as the gravitational force.

To be sure, under normal conditions there are no electron/positron systems in our surroundings, for positrons virtually do not exist. Instead our universe (as far as we know) is held together electromagnetically by electron/proton attractions. The proton is 1836 times as massive as the electron, so that the gravitational attraction is increased without a concomitant increase in electromagnetic attraction. In this case the ratio F_e/F_g is only 2.3×10^{39}.

There are two other major forces in the physical world. There is the nuclear strong interaction force which is over a hundred times as strong as even the electromagnetic force; and the nuclear weak interaction force, which is considerably weaker than the electromagnetic force. All three, however, are far, far stronger than the gravitational force.

In fact, the force of gravity—though it is the first force with which we are acquainted, and though it is always with us, and though it is the one with a strength we most thoroughly appreciate—is *by far the weakest known force in nature*. It is first and rearmost!

What makes the gravitational force *seem* so strong?

First, the two nuclear forces are short-range forces which make themselves felt only over distances about the width of an atomic nucleus. The electromagnetic force and the gravitational force are the only two long-range forces. Of these, the electromagnetic force cancels itself out (with slight and temporary local exceptions) because both an attraction and a repulsion exist.

This leaves gravitational force alone in the field.

What's more, the most conspicuous bodies in the universe happen to be conglomerations of vast mass, and we live on the surface of one of these conglomerations.

Even so, there are hints that give away the real weakness of gravitational force. Your weak muscle can lift a fifty-pound weight with the whole mass of the earth pulling, gravitationally, in the other direction. A toy magnet will lift a pin against the entire counterpull of the earth.

Oh, gravity is weak all right. But let's see if we can dramatize that weakness further.

Suppose that the Earth were an assemblage of nothing but its mass in positrons, while the Sun were an assemblage of nothing but its mass in electrons. The force of attraction between them would be vastly greater than the feeble gravitational force that holds them together now. In fact, in order to reduce the electromagnetic attraction to no more than the present gravitational one, the Earth and Sun would have to be separated by some 33,000,000,000,000,000 light years, or about five million times the diameter of the known universe.†

Or suppose you imagined in the place of the Sun a million tons of electrons (equal to the mass of a very small asteroid). And in the place of the Earth, imagine 3⅓ tons of positrons.††

The electromagnetic attraction between these two insignificant masses, separated by the distance from the Earth to the Sun, would be equal to the gravitational attraction between the colossal masses of those two bodies right now.

In fact, if one could scatter a million tons of electrons on the Sun, and 3⅓ tons of positrons on the Earth, you would double the Sun's attraction for the Earth and alter the nature of Earth's orbit considerably. And if you made it electrons, both on Sun and Earth, so as to introduce a repulsion, you would cancel the gravitational attraction altogether and send old Earth on its way out of the Solar System.

Of course, all this is just paper calculation. The mere fact that electromagnetic forces are as strong as they are means that you cannot collect a significant number of like-charged particles in one place. They would repel each other too strongly.

Suppose you divide the Sun into marble-sized fragments

† Well, a little over *one* million times would be more in accord with the latest findings of distant quasars.

†† When this article first appeared in October 1964, Linus Pauling wrote to say I had made a huge arithmetical error in it. He didn't bother to tell me where and I went through the article in a panic. I found it in this paragraph and the error is corrected in the version you see here.

and strewed them through the Solar System at mutual rest. Could you, by some manmade device, keep those fragments from falling together under the pull of gravity? Well, this is no greater a task than that of getting hold of a million tons of electrons and squeezing them together into a ball.

The same would hold true if you tried to separate a sizable quantity of positive charge from a sizable quantity of negative charge.

If the universe were composed of electrons and positrons as the chief charged particles, the electromagnetic force would make it necessary for them to come together. Since they are anti-particles, one being the precise reverse of the other, they would melt together, cancel each other, and go up in one cosmic flare of gamma rays.

Fortunately, the universe is composed of electrons and protons as the chief charged particles. Though their charges are exact opposites (−1 for the former and +1 for the latter), this is not so of other properties—such as mass, for instance. Electrons and protons are not anti-particles, in other words, and cannot cancel each other.

Their opposite charges, however, set up a strong mutual attraction that cannot, within limits, be gainsaid. An electron and a proton therefore approach closely and then maintain themselves at a wary distance, forming the hydrogen atom.

Individual protons can cling together despite electromagnetic repulsion because of the existence of a very short-range nuclear strong interaction force that sets up an attraction between neighboring protons that far over-balances the electromagnetic repulsion. This makes atoms other than hydrogen possible.

In short: nuclear forces dominate the atomic nucleus; electromagnetic forces dominate the atom itself; and gravitational forces dominate the large astronomic bodies.

The weakness of the gravitational force is a source of frustration to physicists.

The different forces, you see, make themselves felt by transfers of particles. The nuclear strong interaction force, the strongest of all, makes itself evident by transfers of pions (pi-mesons), while the electromagnetic force (next strongest) does it by the transfer of photons. An analogous

NASA

THE EARTH

It is so easy to assume that the Earth is the better part of the Universe and that the rest is all an insignificant addendum. In fact, through most of mankind's stay on the planet that is exactly what the general opinion was.

The Earth is all around us and fills our consciousness. It seems incredibly vast and the heavenly bodies are just small objects in the sky. The gravitational effect of the Earth is all that seems to matter. We know that we and everything else on Earth are drawn downward. The effect of this "gravity" was known to the ancient Greeks.

Newton was the first to see that the Moon was kept in orbit about the Earth by the pull of the Earth's gravitational field; the same pull that drew an apple from the tree down to the ground. This might seem merely to extend the Earthbound force farther, but Newton realized that if the Earth's gravitational field held the Moon in orbit, then the Sun's gravitational field held the Earth in orbit. From this there came the grand generalization—perhaps the greatest single inspiration of the human mind—that *every* body exerted a gravitational effect on *every other* body.

Copernicus had put the Earth in motion about the Sun but did nothing to show that any other body had Earthly properties. Newton showed that all bodies, Earth and heavenly alike, shared in the same gravitational property. That, more than anything else, showed that Earth was only a body among bodies.

Yet however well we might understand this in our minds, and however thoroughly we might inspect a manmade globe, nothing ever put Earth in its place as effectively as the manner in which its photograph as a great globe in the sky did. The picture here was taken by Apollo 16 astronauts and in it you can see the United States through a rift in the clouds.

particle involved in weak interactions (third strongest) has been reported. It is called the "w particle" and as yet the report is a tentative one.*

So far, so good. It seems, then, that if gravitation is a force in the same sense that the others are, it should make itself evident by transfers of particles.

Physicists have given this particle a name, the "graviton." They have even decided on its properties, or lack of properties. It is electrically neutral and without mass. (Because it is without mass, it must travel at an unvarying velocity, that

* *Very* tentative. As of 1974, ten years after this article first appeared, the w particle, while generally accepted, has not actually been detected.

of light.) It is stable, too; that is, left to itself, it will not break down to form other particles.

So far, it is rather like the neutrino, which is also stable, electrically neutral, and massless (hence traveling at the velocity of light).

The graviton and the neutrino differ in some respects, however. The neutrino comes in two varieties, an electron neutrino and a muon (mu-meson) neutrino, each with its anti-particle; so there are, all told, four distinct kinds of neutrinos. The graviton comes in but one variety and is its own anti-particle. There is but one kind of graviton.

Then, too, the graviton has a spin of a type that is assigned the number 2, while the neutrino along with most other subatomic particles have spins of ½. (There are also some mesons with a spin of 0 and the photon with a spin of 1.)

The graviton has not yet been detected.† It is even more elusive than the neutrino. The neutrino, while massless and chargeless, nevertheless has a measurable energy content. Its existence was first suspected, indeed, because it carried off enough energy to make a sizable gap in the bookkeeping.

But gravitons?

Well, remember that factor of 10^{42}!

An individual graviton must be trillions of trillions of trillions of times less energetic than a neutrino. Considering how difficult it was to detect the neutrino, the detection of the graviton is a problem that will *really* test the nuclear physicist.

† In 1969, five years after this article first appeared, Joseph Weber reported the detection of gravitational waves, which is equivalent to detecting the graviton. This finding has not been confirmed, however, and is in serious doubt, so that my last sentence in this article still stands, perhaps.

Six
THE RIGID VACUUM

Probably the greatest dilemma facing the man who wants to write science fiction on the grand scale—and who is also conscientious—is that of squaring the existence of an interstellar society with the fact that travel at velocities greater than that of light in a vacuum (186,200 miles per second)* is considered impossible.

There are a number of ways out, however, and I'll mention three. The most honest is to accept the limitation, and to assume instead that travelers experience time-dilatation. That is, a trip that takes two weeks from their own standpoint may take twenty years from the standpoint of those at home. This, of course, creates difficulties of plotting, and is therefore unpopular among most writers.

The most daring and intriguing solution is E. E. Smith's "inertialess drive," in which matter is assumed to be freed of inertia. (As far as we know, by the way, this is impossible.) Matter without inertia can undergo an acceleration of any size by the application of any force however small. Smith assumed that matter would then be capable of attaining any velocity, even one far beyond that of light.

Actually, this is not so. Photons and neutrinos have zero mass and therefore zero inertia, yet travel no faster than the velocity of light. Consequently, an inertialess drive would be of no help.

There's another flaw here, too. The resistance of even the thinly-spread gas and dust in interstellar space would become significant as velocity rises. Eventually a limit to velocity would be set beyond which one could expect the ship to be melted and its occupants broiled.

The most pedestrian solution is the one I use myself, which is to speak of "hyperspace." This involves higher dimensions and one usually drags in analogies concerning

* I use slightly different figures for the speed of light from time to time. The best current figure, obtained in 1972, is 186,282.3959 miles per second.

one's going through a piece of paper to get to the other side, instead of traveling all the way over to the far distant edge.

In contrast to all this great thought given over to the problem of interstellar travel, very little is devoted to the problem of interstellar communication. According to the relativistic viewpoint of the universe, it is not simply matter that cannot be transported at speeds greater than that of light in a vacuum; it is any form of meaningful symbol.

Well, then, suppose you don't want to travel to Sirius to see your girl friend; suppose you just want to put in a call and speak to her. How do you do that without having to wait sixteen years for the signal to make the round trip.

As far as I know, when this facet of the problem is considered, it is tossed off with the word *sub-etheric*.† And that, at last, brings me to the point. I want to explain what a science-fiction writer means by *sub-etheric*, and I want to do it in my own fashion; i.e. the long way round.

The word *ether* has had a long and splendid history, dating back to the time it was coined by Aristotle about 350 B.C.

To Aristotle the manner in which an object moved was dictated by its own nature. Earthy materials fell and fiery particles rose because earthly materials had an innate tendency to fall and fiery particles an innate tendency to rise. Therefore, since the objects in the heavens seemed to move in a fashion characteristic of themselves (they moved circularly, round and round, instead of vertically, up or down), they had to be made of a substance completely different from any with which we are acquainted down here.

It was impossible to reach the heavens and study this mysterious substance, but it could at least be given a name. (The Greeks were good at making up names, whence the phrase, "The Greeks had a word for it.") The one proprety of the heavenly objects that could be perceived, aside from their peculiar motion, was, however, their blazing luminosity. The sun, moon, planets, stars, comets, and meteors all gave off light. The Greek word for "to blaze" (transliterated into our alphabet) is *aithein*. Aristotle therefore called the

† At least, that's how *I* toss it off.

heavenly material *aither*, signifying "that which blazes." In Aristotle's day it was pronounced "i'ther," with a long *i*.

The Romans adopted this Greek word, because to the Romans, Greek was the language of learning and the average Roman pedant adapted all the Greek words he could, just as our modern pedants are as Latinized as possible, and as the pedant of the future will drag in all the ancient English he can. The Romans transliterated *aither* into *aether*, making use of the diphthong *ae* to keep the pronunciation correct, since that, in the Latin of Cicero's day, was pronounced like a long *i*. (*Caesar* is pronounced "Kaiser," as the Germans know, but we don't.)

The British keep the Latin spelling of *aether*, but Latin (and Greek, too) underwent changes in pronunciation after classical times, and by medieval times *ae* had something of a long *e* sound. So *aether* came to be pronounced "ee'ther."

But if it's going to be pronounced that way, why not get rid of the superfluous *a* and spell it "ether." This, actually, is what Americans do.

(The Greek word for blood is *haima*, and now you can figure out for yourself why we write "hemoglobin" and the British write "haemoglobin.")

This Aristotelian sense of the word *ether* is still with us whenever we speak of something that is heavenly, impalpable, refined of all crass material attributes, incredibly delicate, and so on and so on, as being "ethereal."

By 1700 the Greek scheme of the universe had fallen to pieces. The sun, not the earth, was the center of the planetary system, and the earth moved about the sun, as did the other planets. The motions of the heavenly bodies, including the earth, were dictated solely by gravity; and the force of gravity operated on ordinary objects as well. The laws of motion were the same for all matter and did not in the least depend on the nature of the moving object. Seeming differences were the result of the intrusion of additional effects: buoyancy, friction, and so on.

In the general smashup of Aristotelian physics, however, one thing remained—the ether.

You see, if we wipe out the notion that objects move according to some inner compulsion, then they must move according to some compulsion imposed upon them from outside. This outer compulsion, gravity, bound the earth to the sun, for instance—but, come to think of it, how?

If you wish to exert a force on something; to push it or pull it; you must make contact with it. If you do not make direct contact with it, then you make indirect contact with it; you push it with a stick you hold in your hand or pull it with a hook. Or you can throw a stick (or a boomerang) and the force you impart to the stick is carried, physically, to the object you wish to affect. Even if you knock down a house of cards with a distant wave of the hand, it is still the air you (so to speak) throw at the cards, that physically carries the force to the cards.

In short, something physical must connect the object forcing and the object forced. Failing that, you have "action at a distance," which is a hard thing to grasp and which philosophers of science seem to be reluctant to accept if they can think of any other way out of a dilemma.

But gravity seems to involve action at a distance. Between the sun and the earth, or between the earth and the moon, is a long stretch of nothing, not even air. The force of gravitation makes itself felt across the vacuum; it is therefore conducted across it; and the question arises: What does the conducting? What carries the force from the sun to the earth?

The answer consisted of Aristotle's word again, *ether*. This new ether, however, was not something that made up the heavenly bodies. The seventeenth-century scientist rather suspected the heavenly bodies were made up of ordinary earthly matter. Instead, ether was now viewed as making up the apparently empty volume through which all these bodies of matter moved. In short, it made up space; it was, so to speak, the very fabric of space.

Exactly what ether's properties were could not be shown by direct observation, for it could not be directly observed. It was not matter or energy, for when only ether was present, what seemed to exist to our senses and to our measurements was a vacuum—*nothing*. At the same time, ether (whatever it was) was to be found not only in empty space but permeating all matter, too, for the conduction of the gravitational force did not seem to be interfered with by matter. If, as during a solar eclipse, the moon passed between the earth and the sun, the earth's movements were not affected by a hair. The force of gravity clearly traveled, unchanged and undiminished, through two thousand miles of matter. Consequently, the ether permeated the moon and, by a reasonable generalization, it permeated all matter.

Furthermore, ether did not interfere with the motion of the planets. Planets moved through the ether as though it were not there. Matter and ether, then, simply did not interact at all. Ether could conduct forces but was not itself subject to them.

This meant that ether was not moving. How could it move unless some force were applied to it, and how could such a force be exerted upon it if matter would not interact with it? Or, to put it another way, ether is indistinguishable from a vacuum, and can you picture a way in which you can exert force on a vacuum (not on a container which may hold a vacuum, but on the vacuum itself) so as to impart motion to it?

This was an important point. As long as astronomers were sure that the earth was the motionless center of the universe (even if it rotated, the *center* of the earth was motionless), it was possible to work up laws of motion with confidence. Motion was a concept that meant something. If the earth traveled about the sun, however, then while you were working out the laws of motion relative to the earth, you would be plagued by wondering whether those laws would make sense if the same motion were viewed relative to Mars, for instance.

Actually, if one could find something that was at rest and refer motion to that, then the laws of motion would still make sense because the earth's motion with reference to the something at rest could be subtracted from the object's motion with reference to the something at rest and that would leave the object's motion with reference to the earth, and the laws would still apply and you wouldn't have to worry about the motions with reference to Mars or to Alpha Centauri or anything else.

And this was where the ether came in. Ether could not move; motion was alien to the very concept of ether; so it could be considered in a state of Absolute Rest. This meant there was such a thing as Absolute Motion, since any motion could, in principle, be referred to the ether. The framework of space and time within which such absoluteness of rest and motion can exist can be referred to as Absolute Space and Absolute Time.

A century after Newton, ether was to be called upon again. The force of gravity, after all, was not the only entity to reach us across the stretches of empty space; another entity was light.

Light did not, however, raise the anxiety at first that gravitation did, for it did not act as gravity did. For one thing, light could be shielded. When the moon interposed itself between ourselves and the sun, light was cut off even though gravity wasn't. Thin layers of matter could completely block even strong light, so that it would seem that light could not be conducted by the ether which permeated all matter.

Furthermore, the direction in which a light ray traveled could be charged ("refracted") by passing it from one medium to another, as from air to water, although ether permeated both media equally. The direction in which gravity exerted its force could not be changed by any known method.

Newton postulated, therefore, that light consisted of tiny particles moving at great velocities. In this way, light required no ether and yet did not represent action at a distance either, for the effect was carried across a vacuum, physically, by moving objects. Furthermore, the particle theory could be easily elaborated to explain the straight-line motion of light, and its ability to be reflected and refracted.

There were opposing views in Newton's time to the effect that light was a wave form, but this made no headway. The wave forms then known (water waves and sound waves, for instance) did not travel in straight lines but easily bent around obstacles. This was not at all the way light acted and therefore light could not be a wave form.

In 1801, however, an English physician, Thomas Young, showed that it was possible to combine two rays of light in such a way as to get alternating bands of light and darkness ("interference fringes"). This seemed difficult to explain if light consisted of particles (for how could two particles add together to make no particles?), but very easy to explain if light were a wave form. Suppose the wave of one light ray were on its way up and the wave of the other were on its way down. The two effects would cancel, for no net motion at all, and there would be darkness.

Furthermore, it could be shown that a wave form would move about obstacles that were of a size comparable to its own wavelength. Obstacles larger than that would be increasingly efficient (as their size increased) in reflecting the wave form. Where obstacles vastly larger than the wavelength were concerned, the wave form would seem to travel in straight lines and cast sharp shadows.

Well, ordinary sound waves have wavelengths measured in feet and yards. Young, however, was able to deduce the wavelength of light from the width of the interference fringes and found it to be something like a sixty-thousandth of an inch. As far as obstacles of ordinary size were concerned, obstacles large enough to see, light traveled in straight lines and cast sharp shadows even though it was a wave form.

But this view did not take over without opposition. It raised serious philosophical problems. It makes one ask at once: "If light consists of waves, then what is waving?" In the case of water waves, water molecules are moving up and down. In the case of sound waves through air, air molecules are moving to and fro. But light waves?

The answer was forced upon physicists. Light can travel through a vacuum with the greatest ease, and the vacuum contained nothing but ether. If light was a wave form, it had to consist, therefore, of waves of ether.

But then how account for the fact that light could be reflected, refracted, and absorbed, when gravitation carried by the same ether could not? Was it possible that there were two ethers with different properties, one to conduct gravity and one to conduct light? The question was never answered, but through the nineteenth century, light was far more crucial to the development of theoretical physics than gravity was, and it was the particular ether that carried light that was under continual discussion. Physicists referred to it as the "luminiferous ether" (Latin for "light-carrying ether").

But difficulties were to arise in the case of the luminiferous ether that never arose in the case of the gravity-carrying ether. You see, there are two kinds of wave forms—

In water waves, while the wave motion itself is progressing, let us say, from right to left, the individual water molecules are moving up and down. The movement of the oscillating parts is in a direction at right angles to the movement of the wave itself. This type of wave form, resembling a wriggling snake, is a "transverse wave."

In sound waves the individual molecules are moving back and forth in the same direction that the sound wave is traveling. Such a wave form (a bit harder to picture) is a "longitudinal wave."

Well, then, what kind of a wave is a light wave, transverse

The Granger Collection

THOMAS YOUNG

Young, born in Milverton, England, on June 13, 1773, was an infant prodigy. He could read at two and had worked his way twice through the Bible at four. During his youth, he learned a dozen languages, including not only Greek, Latin, and Hebrew, but also Arabic, Persian, Turkish, and Ethiopian. He could also play a variety of musical instruments, including the bagpipes. The best of it was that he grew up to be an adult prodigy who was called "Phenomenon Young" at Cambridge. Then to add to all this, a rich uncle died in 1797 and left Young financially independent.

He took up medicine and obtained his degree at Göttingen, Germany, in 1796. As a physician he was unsuccessful because he lacked a suave bedside manner, but in research he was first class. While still a medical student, he was the first to discover the manner in which the lens of the eye changes shape in focusing on objects at different distances. In 1801, he described the reason for astigmatism—irregularities of the curvature of the cornea.

His greatest contribution to science, however, was his demonstration of the wave nature of light when he showed that two beams of light could interfere, producing alternate bands of light and darkness. On a smaller scale, he was the first, in 1807, to use the word "energy" in its modern sense.

As though this varied activity were not enough, Young contributed many and varied articles to the Encyclopedia Britannica, then reached beyond the natural sciences altogether and took up the problem of the ancient hieroglyphic language of the Egyptians in 1814. He gave up his medical practice to tackle this linguistic problem and was the first to make progress in this direction.

He died in London, on May 10, 1829, a month short of his fifty-sixth birthday.

or longitudinal? At first, everyone voted for longitudinal waves—even Young did—for reasons I'll shortly explain.

Unfortunately, one annoying fact intervened. Back in Newton's time a Dutch physician, Erasmus Bartholin, had discovered that a ray of light, upon entering a transparent crystal of a mineral called Iceland spar, was split into two rays. The separation was brought about because the original ray was bent by two different amounts. Everything seen through Iceland spar seemed double, and the phenomenon was called "double refraction."

In order for a ray of light to bend in two different directions on entering Iceland spar, the components of light had to exist in two different varieties, or, if there were only one variety, that variety had to show some sort of asymmetry.

Newton tried to adjust the particle theory of light to account for this, and made a heroic effort, too. Through sheer intuition, he caught a glimmer of our modern view of light as consisting of both particles and waves, two centuries ahead of time. However, after Newton's death, the lesser minds that followed him thought of a much better way of accounting for "double refraction." They ignored it.

What about the wave theory? Well, no one could think of a way to make a longitudinal wave explain double refraction, but transverse waves were another matter.

Imagine that your eye is a piece of Iceland spar and that a ray of light is coming directly toward it. The ether, as was then supposed, would be undulating at right angles to the direction of motion, but there are an infinite number of directions that would be at right angles to the direction of motion. As the light comes toward you, the ether could be moving up and down, or right and left, or diagonally (turned either clockwise or counterclockwise), to any extent.

Every diagonal undulation can be divided into two components, a vertical one and a horizontal one, so in the last analysis we can say that the light ray approaching us is made up of vertical undulations and horizontal undulations. Well, Iceland spar can choose between them. The vertical undulations bend to one extent, the horizontal to another, and where one ray of light enters, two emerge.

It is a good question as to why Iceland spar should do this and not glass, but the question is not pertinent to this discussion and I shall leave that to another essay some day. What does matter is simply that longitudinal waves could not be used to explain double refraction and transverse waves could and the conclusion had to be, then, that light consisted of transverse waves. The theory of light as a transverse wave form was worked out in the 1820s by a French physicist named Augustin Jean Fresnel.

This aroused a furor indeed, for the manner in which longitudinal waves and transverse waves are conducted show important differences. Longitudinal waves can be conducted by matter in any state, gaseous, liquid, or solid. Thus, sound waves travel through air, through water, and through iron with equal ease. If light were a longitudinal wave, then the luminiferous ether could be viewed as an exceedingly subtle gas; so subtle as to be indistinguishable from a vacuum. It would still be capable, in principle, of conducting light.

Transverse waves are more particular. They cannot travel through the body of a gas or liquid. (Water waves agitate the surface of water, but cannot travel through the water itself.) Transverse waves can travel through solids only. This means that if the luminiferous ether conducts light, and if light is a transverse wave, then the luminiferous ether must have the properties of a solid!

And there is worse to follow. For atoms or molecules to engage in periodic motion (as they must, to establish a wave form), they must have elasticity. They must spring back into position, if deformed out of it, overshoot the mark, spring back again, overshoot the mark again, and so on. The speed with which an atom or molecule springs back into position depends upon the rigidity of the material. The more rigid, the faster the snapback, the faster the oscillation as a whole, and the faster the progress of the wave form. Thus sound waves progress more rapidly through water than through air, and more rapidly through steel than through water.

It works in reverse. If we know the velocity at which a wave form travels through a medium, we can calculate how rigid it must be.

Well, what is the velocity of light through a vacuum; i.e. through the ether? It is 186,200 miles per second, and this was known in Fresnel's time. For transverse waves to travel that rapidly, the conducting medium must be rigid indeed—more rigid than steel.

And so there's the picture of the luminiferous ether; a substance indistinguishable from a vacuum yet more rigid than steel. A rigid vacuum! No wonder physicists tore their hair.

A generation of mathematicians worked out theories to account for this wedding of the mutually exclusive and managed to cover the general inconceivability of a rigid vacuum with a glistening layer of fast-talking plausibility. As for an actual physical picture of the luminiferous ether, the best that could be advanced was that it was a substance something like the modern Silly Putty. It yielded freely to a stress applied relatively slowly (as by a planet moving at

The Granger Collection

HEINRICH RUDOLF HERTZ

Hertz, the son of a Jewish lawyer, was born in Hamburg, Germany, on February 22, 1857. He started his studies in engineering but abandoned that for physics and obtained his Ph.D., *magna cum laude*, in 1880.

He was interested in Maxwell's equations that predicted there could be light-like radiations with wavelengths much shorter, or much longer than those of light itself. In 1888, he set up an electrical circuit that oscillated, surging into first one, then another, of two metal balls separated by an air gap. Each time the potential reached a peak in one direction or the other, it sent a spark across the gap.

With such an oscillating spark, radiation should be produced with a wavelength equal to the distance light could travel in the time between alternate surges. Such wavelengths would be far greater than those of light.

To detect such long-wave radiation, Hertz used a simple loop of wire with a small air gap at one point. Just as current gave rise to radiation in the first coil, so the radiation ought to give rise to a current in the second coil. Sure enough, Hertz was able to detect small sparks jumping across the gap in his detector coil. By moving the detector coil to various points in the room, Hertz could tell the shape of the waves by the intensity of spark formation. These "Hertzian waves," which Hertz showed to be an "electromagnetic radiation" like light, are now called radio waves.

Radio waves led to a tremendous advance in technology —radio transmission, television, radar, masers, radio astronomy, and so on, but Hertz did not live to see even the beginning of it. He died in Bonn, Germany, after a long illness due to chronic blood poisoning, on January 1, 1894, before his thirty-seventh birthday.

two to twenty miles per second), but rigidly resisted a stress applied rapidly (as by light traveling at 186,200 miles a second).

Even so, physicists would undoubtedly have given up the ether in despair if it weren't so useful as the only way to avoid action at a distance. And instead of growing less useful with time, it grew more so, thanks to the work of the Scottish mathematician James Clerk Maxwell. This came about as follows.

Long before Newton had worked out the theory of gravitation, two other types of action-at-a-distance forces were known: magnetism and static electricity. Both attracted objects even across a vacuum and both types of forces, it therefore seemed, had to be conducted by the ether. (In fact, before the theory of gravitation had been put forth, men such as Galileo and Kepler speculated that magnetic forces must bind planets to the sun.)

But there again—was there a separate ether for magnetism and one for electricity, as well as one for light and one for gravity? Were there four ethers altogether, each with its own properties? If so, things were worse than ever. This piling up of four different vacuums, one as rigid as steel, and the other three who-knows-what, threatened to rear a structure that would topple under its own weight and bury the edifice of physics in its ruins.

In the mid-nineteenth century, Maxwell subjected the matter to acute mathematical analysis and showed that he could build up a consistent picture of what was known of electricity and magnetism, and in so doing, maintained that the two forces were interrelated in such a way that one could not exist without the other. There was neither electricity nor magnetism, but "electromagnetism."

Furthermore, if an electrically charged particle oscillated, it radiated energy in the form of a wave, with a frequency equal to that of the oscillation period. In other words, if the charge oscillated a thousand times a second, a thousand waves were formed each second. The velocity of such a wave worked out to a certain ratio which, once solved, turned out to be just about exactly the speed of light.

Maxwell could not believe this to be a coincidence. Light, he insisted, was an "electromagnetic radiation." (Light has a frequency of several hundred trillion waves per second, and where was the electric charge that oscillated at such a rate? Maxwell couldn't answer that, but a generation later, the electrons within the atom were discovered and the question was answered.)

Such a theory was delightful. It unified electricity, magnetism, and light into different aspects of one phenomenon and made one ether do for all three.†† This simplified the ether concept and made it explain much more than before. (At this point, it should perhaps have been renamed the "electromagniferous ether," but it wasn't.) If Maxwell's theory held up, physicists could grow much more comfortable with the ether concept. But would it hold up?

One way to establish a theory is to make predictions based upon its tenets and have them turn out to be so.

†† This leaves gravity out, but all efforts to join gravity to electromagnetism as a fourth aspect (a "unified field theory") have failed. Einstein devoted half his life to it but did not succeed.

To Maxwell, it seemed that since electric charges could oscillate in any period, there should be a whole family of electromagnetic radiations with frequencies greater than those of light and smaller than those of light, and to all degrees.

This prediction was borne out in 1888 (after Maxwell's too-early death, unfortunately) when the German physicist Heinrich Hertz managed to get an electric current to oscillate not very rapidly and then detected very low-frequency electromagnetic radiation. This low-frequency, long-wavelength radiation is what we now call "radio waves."

Radio waves, being electromagnetic radiation, are conducted through the ether at 186,200 miles per second. This is the limiting speed of communication by any form of electromagnetic radiation.

But if we grant the ether concept, suppose we imagine a "sub-ether," one that permeates the ether itself as ether permeates matter, and one that has all the properties of ether greatly intensified. It would be even more tenuous and undetectable and at the same time far more rigid. It would, in other words, be a super-rigid super-vacuum. It might even be conjectured that gravitational force, still unaccounted for by Maxwell's theory, would travel through such a sub-ether.

In that case, wave forms (perhaps gravitic, rather than electromagnetic) would travel through it at far greater velocities than that of light. The stars of the galactic empire might then not be too far apart for rapid communication.

And there is your word *sub-etheric*.

Now isn't that an exciting idea? Might it not even be valid? After all, if the ether concept is granted . . .

Ah, but is it granted?

You see, Hertz's discovery of waves that confirmed Maxwell's electromagnetic theories and seemed to establish the ether concept once and for all, had come too late. Few realized it at the time, but the year before Hertz's discovery, the ether concept had been shattered past retrieval.

It happened through one little experiment that didn't work.

And if you read the next chapter, you'll learn about it.

Seven
THE LIGHT THAT FAILED

In the summer of 1962 a fetching young lady from *Newsweek* asked permission to interview me; permission which I granted at once, you may be sure. It seems that *Newsweek* was planning to do a special issue on the space age, and it was this young lady's job to gather some comments on the matter by various science-fiction personalities.

I discoursed learnedly on science fiction to her, filling ninety unforgiving minutes with sixty seconds' worth of distance run each, before I ungripped her with my glittering eye.

Eventually, the special issue appeared, dated October 8, 1962, and there, on page 104, was three quarters of a page devoted to science fiction (and not bad commentary either; no complaints on that score). Within that section, every bit of my long-winded brilliance was discarded with the exception of one remark which read as follows:

"But sci-fi is 'a topical fairy tale where all scientists' experiments succeed,' comments Isaac Asimov. . . ."*

Ever since then, this quotation has bothered me. Oh, I said it; I wasn't misquoted. It's just that I seem to have implied that what scientists want are experiments that succeed, and that is not necessarily true.

Under the proper circumstances, a failure, if unexpected and significant, can do more for the development of science than a hundred routine successes. In fact, the most dramatic single experiment in the last three and a half centuries was an outright failure of so thunderous a nature as to help win its perpetrator a Nobel Prize. What a happy fairy tale for scientists it would be if all experiments failed like that!

This fits in, fortunately, with the fact that in the previous chapter, I discussed the ether and ended at a point where it seemed enthroned immovably in the very fabric of physics. I said then that at the very peak of its power and

* My natural modesty forbids my quoting the rest of the passage, but you can look it up for yourself if you like. Please do.

prosperity, the ether was dashed from its throne and destroyed.

The man who brought about that destruction was an American physicist named Albert Abraham Michelson. What started him on the track was a peculiar scientific monomania; Michelson got his kicks out of measuring the velocity of light. Such a measurement was his first scientific achievement, and his last, and just about everything he did in science in between grew out of his perpetual efforts to improve his measurements.

And if you think I'm going to go one step farther without retreating three centuries to discuss the history of the measurement of the velocity of light, you little know me.

Throughout ancient and medieval times the velocity of light was assumed (by those who thought about the matter at all) to be infinite. The Italian scientist Galileo was the first to question this. About 1630 he proposed a method for measuring the velocity of light.

Two people, he suggested, were to stand on hilltops a mile apart, both carrying shielded lanterns. One was to uncover his lantern. The other upon seeing the light was at once to uncover his own lantern. If the first man measured the time that elapsed between his own uncovering and the sight of the spark of light from the other hill, he would know how long it took light to cover the round distance. Galileo had actually tried the experiment, he said, but had achieved no reasonable results.

It is not hard to see why he had failed, in the light of later knowledge. Light travels so quickly that the time lapse between emission and return was far too short for Galileo to measure with any instrument that then existed. There would be a small time lapse to be sure, but that represented the time it took for the assistant to think "Hey, there's the old man's light" and get his own light uncovered.

All that Galileo could possibly have shown by his experiment, which was correct in principle, was that if the velocity of light was not infinite, then it was at least very, very fast by ordinary standards. Still, it was useful to show even this much.

The next step was taken nearly half a century later. In 1676 a Danish astronomer Olaus Roemer was working at the Paris Observatory, observing Jupiter's satellites. Their times of revolution had been carefully measured, so it

The Bettmann Archive, Inc.

THE TELESCOPE

The telescope (the one shown here is that in Greenwich, England) produced a more dramatic change in man's scientific and *cultural* history than any other scientific instrument I can think of.

When Galileo looked at the Moon with a telescope and saw mountains, craters, and "seas," that was the final piece of evidence in favor of a plurality of worlds. The

Earth was not the only object on which life could conceivably exist.

Then when Galileo, with his telescope, discovered four satellites circling Jupiter, that was the first evidence that the Earth was not necessarily the center of all heavenly turnings. When he and his telescope discovered that the planet Venus went through the complete cycle or phases as the Moon did, something that Copernicus' system with the Sun at the center predicted, while the older Earth-centered view did not, that settled the matter in favor of Copernicus.

Mankind might have argued over the respective merits of Earth-as-center and Sun-as-center forever without coming to any decision, save for the telescope. And with Earth dethroned as center, man had to change his opinion of himself, and of his role in the Universe. The great man-centered drama of sin-and-redemption, constructed in earlier times, looked puny against the new Universe.

Galileo and his telescope also detected myriads of stars too faint to see with the unaided eye and that was the first indication that the Universe beyond the planets was far larger than anyone had imagined.

In the nearly four centuries since Galileo, the telescope has continued to astonish mankind. Even in our own generation, the descendants of that early telescope have helped bring to our attention quasars, pulsars, and black holes and have made the Universe still larger, still stranger, and still more violent than we had imagined.

seemed possible to predict the exact moments at which each would pass into eclipse behind Jupiter, and this, too, had been done.

To Roemer's surprise, however, the moons were being eclipsed at the wrong times. At those times of the year when the earth was approaching Jupiter, the eclipses came more and more ahead of schedule, while when earth was receding from Jupiter, they fell progressively further behind schedule.

Roemer reasoned that he did not see an eclipse when it took place, but only when the cut-off end of the light beam reached him.The eclipse itself took place at the scheduled moment, but when the earth was closer than average to Jupiter, he *saw* the eclipse sooner than if the earth were farther than average from Jupiter. Earth was at a minimum distance from Jupiter when both planets were in a line on the same side of the sun, and it was at a maximum distance when both planets were in a line on opposite sides of the sun. The difference between those distances was exactly the diameter of the earth's orbit.

The difference in time between the earliest eclipse of the satellites and the latest eclipse must therefore represent the time it took light to travel the diameter of earth's orbit. Roemer measured this time as twenty-two minutes and, accepting the best figure then available for the diameter of the earth's orbit, calculated that light traveled at a velocity of 138,000 miles per second. This is only three quarters of what is now accepted as the correct value,† but it hit the correct order of magnitude and for a first attempt that was magnificent.

Roemer announced his results and it made a small splash but aroused as much opposition as approval and the matter was forgotten for another half century.

In the 1720s the English astronomer James Bradley was hot on the trail of the parallax of the stars. This had become a prime astronomical problem after Copernicus had first introduced the heliocentric theory of the solar system. If the earth really moved about the sun, said the anti-Copernicans, then the nearby stars should seem to shift position ("parallactic displacement") when compared with the more distant stars. Since no such shift was observed, Copernicus must be wrong.

"Ah," said the Copernicians in rebuttal, "but even the nearest stars are so distant that the parallactic displacement is too small to measure."

Yet even after astronomers had all adopted the heliocentric theory, there was still discomfort over the question of the stellar parallax. This business of "too small to measure" seemed very much like an evasion. Observation should be refined to the point where the shift *could* be measured. That would accomplish two things. It would

† The actual maximum difference of eclipse times, by later measurements, turned out to be sixteen minutes thirty-six seconds. The diameter of the earth's orbit is about 185,500,000 miles, and I leave it to you, O Gentle Reader, to calculate a good approximation of the correct velocity.

show how far the nearest stars were, and it would be the final proof that the earth was moving around the sun.

Bradley's close observations did, indeed, demonstrate that the stars showed a tiny displacement through the year. However, this displacement was not of the right sort to be explained by the earth's motion. Something else had to be responsible and it was not until 1728 that a suitable explanation occurred to Bradley.

Suppose we consider the starlight bombarding the earth to be like rain drops falling in a dead calm. If a man were standing motionless in such a rainstorm, he would have to hold an umbrella vertically overhead to ward off the vertically-dropping rain. If he were to walk, however, he would be walking into the rain and he would have to angle the umbrella forward, or some drops that would just miss the umbrella would nevertheless hit him. The faster he walks, the greater the angle at which he would have to tilt his umbrella.

In the same way, to observe light from a moving earth, the telescope has to be angled very slightly. As the earth changes the direction of its motion in its course about the sun, the slight angle of the telescope must be changed constantly and the star seems to mark out a tiny ellipse against the sky (with reference to the sun). Bradley had discovered what is called the "aberration of light."

This was not parallactic displacement and it did not help determine the distance of any stars (that had to wait still another century). Just the same it did prove the earth was moving with respect to the stars, for if the earth were motionless, the telescope would not have to be tilted at all, and the star would not seem to move.

It gave additional information, too. The amount of the aberration of light depended on two factors: the velocity of light and the velocity of the earth's motion in its orbit. The latter was known (about 18½ miles per second); therefore the former could be calculated. Bradley's estimate was that light had a velocity of about 188,500 miles per second. This was only 1.2 percent above the true value.

Two independent astronomical methods had yielded figures for the velocity of light and improved observations showed the two methods yielded roughly the same answer. Was there no way, however, in which the velocity could be measured on earth, under conditions controlled by the experimenter?

The answer was yes, but the world had to wait a century and a quarter after Bradley's discovery before a method was found. The discoverer was a French physicist Armand Hippolyte Louis Fizeau who returned to Galileo's method but eliminated the personal element. Instead of having an assistant return a second beam of light, he had a mirror reflect the first one.

In 1849 Fizeau set up a rapidly turning toothed disk on one hilltop and a mirror on another, five miles away. Light passed through one gap between the teeth of the turning disk and was reflected by the mirror. If the disk turned at the proper speed, the reflected light returned just as the next gap moved into line.

From the velocity at which the wheel had to be turned in order for the returning light to be seen, it was possible to calculate the time it took light to cover the ten-mile-round distance. The value so determined was not as good as those the best astronomic measurements had provided; it was 5 percent off, but it was excellent for a first laboratory attempt.

In 1850 Fizeau's assistant Jean Bernard Leon Foucault improved the method by using two mirrors, one of which revolved rapidly. The revolving mirror reflected the light at an angle and from the angle, the velocity of light could be calculated. By 1862 he had obtained values within a percent or so of the true one.

Foucault went further. He measured the velocity of light through water and other transparent media (you could do this with laboratory methods but not with astronomical methods). He discovered in this way that light moved more slowly in water than in air.

This was important. If the particle theory of light were true, light should move more rapidly in water than in air; if the wave theory of light were true, light should move more slowly in water than in air. By the mid-nineteenth century, to be sure, most physicists had accepted the wave theory. Nevertheless, Foucault's experiment was widely interpreted as having placed the final nail in the coffin of the particle theory.

And now we come to Michelson. Michelson had been born in 1852 in a section of Poland that at that time was under German rule, and he was brought to the United

States two years later. His family did not follow the usual pattern of settling in one of the large East Coast cities. Instead, the Michelsons made their way out to the far West, a region which the forty-niners had just ripped wide open.

The Michelson family did well there (as merchants, not as gold-miners), and young Albert applied for entrance into Annapolis in 1869. He passed the necessary tests, but the son of a war veteran (Civil War, of course) took precedence. It took the personal intervention of President Grant (with an assist from the Nevada congressman who pointed out the political usefulness of such a gesture to the family of a prominent Jewish merchant of the new West) to get Albert in.

He graduated in 1873 and served as a science instructor at the Academy during the latter part of that decade. In 1878 the velocity-of-light bug bit him and he never recovered. Using Foucault's method but adding some ingenious improvements, he made his first report on the velocity, which he announced as 186,508 miles per second. He was approximately 300 miles per second too high, but his result was within one sixth of 1 percent of the actual figure.

In 1882 he tried again, after some years spent in studying optics in Germany and France. This time he came out with a figure of 186,320 miles per second, which was 120 miles per second high, or only one fifteenth of 1 percent off.

Meanwhile, though, it had occurred to Michelson that speeding light could be made to reveal some fundamental secrets of the universe.

One of the big things in the 1880s was the "luminiferous ether" (see the previous chapter). The ether was considered to be motionless, at "absolute rest," and if light were ether waves, then its velocity, if measured carefully enough under appropriate conditions, could give the value of the absolute velocity of the earth; not just its velocity with respect to the sun, but with respect to the very fabric of the universe. Such a value would be of the utmost importance to the philosophy of science, since without it one could never be sure of the validity of all the laws of mechanics that had been worked out since the time of Galileo.

Let me explain how this works. Suppose an airplane were moving at 150 miles an hour and encountered winds of 145 miles an hour. If it were traveling with the wind, the plane

The Granger Collection

ALBERT ABRAHAM MICHELSON

Though Michelson is best known for the Michelson-Morley experiment, he went on thereafter to do many other interesting and important things. He used the interferometer, not only for that experiment, but also to determine the width of heavenly objects by comparing the light rays from both sides and studying the interference fringes produced. In 1920, with a twenty-foot interferometer attached to the new 100-inch telescope, he was able to measure the diameter of the giant star Betelgeuse. It was the first time the diameter of any star was ever measured and the event made the front pages of the New York *Times*.

In 1923, Michelson returned to the problem of the accurate measurement of the velocity of light. In the California mountains he surveyed a 22-mile pathway between two mountain peaks to an accuracy of less than an inch. He made use of a special, eight-sided revolving mirror to reflect light back and forth and by 1927 set the value of the speed of light at 299,798 kilometers per second.

He then tried again, this time using a long tube that could be evacuated so that the speed of light could be measured in a vacuum. Light was reflected and re-reflected until it traveled ten miles in that vacuum. Michelson did not live to make the final measurements, dying in Pasadena on May 9, 1931, but in 1933 the figure was announced as 299,774 kilometers per second.

Another dramatic observation was his microscopic study of the water level in an iron pipe. The tidal changes in level amounted to 4 microns (less than a six-millionth of an inch) but from this it was possible to calculate the intensity of the attraction of the Sun and Moon on the Earth. He showed that the solid earth rose and fell, too, under tidal pull, by a little over a foot.

would seem to move at 295 miles an hour (as viewed from the ground). If it traveled against the wind, it would travel only five miles an hour with respect to the ground. If the velocity of the plane on a windless day were known, then from the difference in velocity produced by the wind, the wind's own velocity could be calculated.

Now suppose the earth were moving through the stationary ether. From a mechanical standpoint, this would be equivalent to the earth standing still while the ether moved past it. Let's take the latter view, for an "ether wind" is easy to visualize.

Light, consisting (as was thought) of ether waves, would move—relative to the earth—with the ether, moving faster than average in the direction of the ether wind, more slowly than average against that direction, and at intermediate velocities in intermediate directions.

Clearly, the velocity of the ether wind could not be very great. If it were blowing at a considerable fraction of the velocity of light, then all sorts of strange phenomena would be observable. For instance, light would radiate outward in an egg-shaped curve instead of in a circle. The fact that no such phenomena were ever observed, meant that the effects must be very small and that earth's absolute velocity could only be a small fraction of the velocity of light.

Michelson turned his attention toward the possibility of measuring that small fraction.

In 1881 Michelson had constructed an "interferometer," a device which was designed to split a light beam in two, send the parts along different paths at right angles to each other, then bring them back together again.

The two rays of light were made to travel exactly the same distance in the process of going and returning, and, therefore, presumably spent the same time on their travels. On returning to their starting point they would merge into one beam again, just as though they had never separated. The merged beams would then display no properties that the original beam had not had.

If, on the other hand, the two light rays had been on their travels for different times, the wave forms of the two rays of light would no longer match; upon merging they would find themselves out of step. There would be places where the waves of one light ray would be moving up while the waves of the other would be moving down. There would then be mutual cancellation ("interference"), and darkness would be produced. The areas of darkness would recur periodically and take the form of a kind of zebra-stripe arrangement ("interference fringes").

The idea was to adjust the instrument so that, as far as was humanly possible, the two light rays would be made to travel just the same distance, and if they then spent the same time at their travels, they would match on merging and no interference fringes would appear.

However, this did not allow for the ether wind, the existence of which was then assumed. If one of the light rays went with the wind, it would return against the wind. The other light ray, sent out at right angles, would then go cross-wind and return cross-wind. It can be shown that, in this case, the time taken by the one ray of light to travel with the wind, then return against it, is slightly longer than the time taken by the other ray of light to travel cross-wind both ways.

The stronger the ether wind, the greater the discrepancy in time; and the greater the discrepancy in time, the wider the interference fringes. By observing the interference fringes, then, Michelson would be able to measure the velocity of the ether wind, and that would give the earth's absolute motion.

Michelson tried the experiment first in Germany, anchoring his interferometer to rock, and driving himself mad trying to eliminate the vibrations set up by city traffic. When he finally sent his split beams of light on their separate paths, he found they brought back no information. The light had failed, and failed miserably. It brought back nothing; no interference fringes at all.

Something had gone wrong, but Michelson did not know what. He let the matter go for a few years.

He returned to the United States, resigned from the Navy, joined the faculty of a new school called Case School of Applied Science, in Cleveland, and there met a chemist named Edward Williams Morley. Morley's ambition had been to be a minister and he took his chemist's job only on condition that he could preach in the school's chapel. His own piece of scientific monomania lay in comparing the atomic weights of oxygen and hydrogen.

Michelson and Morley discussed the interferometer experiment and finally, in 1886, joined forces in order to try again under conditions of the most heroic precautions. They dug down to bedrock to anchor the equipment to the solid planet itself. They built a brick base on which they placed a cement top with a doughnut shaped depression. They placed mercury in the depression and let a wooden float rest upon the mercury. On the wood was a stone base in which the parts of the interferometer were firmly fixed. All was so well balanced that the lightest touch would make the interferometer revolve steadily on its mercury support.

Now they were ready for what was to come to be known as the "Michelson-Morley experiment." Once again a ray of light was split and sent out on its errand; and once again the light failed and brought back nothing. The only interference fringes that were to be seen were tiny ones that clearly represented unavoidable imperfections of the instrument.

Of course, it might be that the rays of light weren't heading exactly upwind and downwind, but in such directions that the ether wind had no effect. However, the instrument could be rotated. Michelson and Morley took measurements at all angles—surely the ether wind had to be blowing in one of those directions. They did even better than that. They kept taking measurements all year while the earth itself changed direction of motion constantly as it moved in its orbit about the sun.

They made thousands of observations, and by July 1887 they were ready to report. The results were negative. They had tried to measure earth's absolute velocity and they had failed and that was that.

There had to be an explanation of this failure and no less than five of them can be considered for a moment. I'll list them.

1) The experiment can be dismissed. Perhaps something was wrong in the equipment or the procedure or the reasoning behind it. Men such as the English scientists Lord Kelvin and Oliver Lodge took that point of view. However, this point of view is not tenable. Since 1887, numerous physicists have repeated the Michelson-Morley experiment with greater and greater precision. In 1960 masers (atomic clocks) were used for the purpose and an accuracy of one part in a trillion was achieved. But always, down to and including the 1960 experiments, the Michelson-Morley failure was repeated. There were no interference fringes. The light rays took precisely the same time to travel in any direction, regardless of the ether wind.

2) Well, then, the experiment is valid and shows there is no ether wind, for any one of four different reasons:

a) The earth is not moving; it is the motionless center of the universe. This would involve so many other paradoxes and would fly in the face of so much astronomical and physical knowledge gained since the time of Copernicus that no scientist seriously advanced this for a moment. However, a friend of mine has pointed out that the only way of disproving this suggestion beyond a doubt is to run the Michelson-Morley experiment elsewhere than on the earth. Perhaps when we reach the moon, we ought to make it one of the early orders of business to repeat the Michelson-Morley experiment there. If it proves negative (and I'm sure it will!), we can certainly conclude that the earth and the moon can't both simultaneously be motionless. That one or the other is motionless is, at least, conceivable; that both are, is not.

b) The earth does move, but in doing so it drags the neighboring ether with it, so that it seems motionless compared with the ether that is right at the surface of the Earth. For that reason, no interference fringes are produced. The

English physicist George Gabriel Stokes suggested this. Unfortunately, this implies that there is friction between the earth and the ether and this would raise the serious question as to why the motion of heavenly bodies wasn't continually being slowed by their passage through the ether. It was as hard to believe in the "ether drag" as in the motionless earth, and Stokes's notion died a quick death.

Two suggestions survived, however;

c) The Irish physicist George Francis FitzGerald suggested that all objects (and therefore all measuring instruments) grew shorter in the direction of motion, according to a formula which was easily derived. This is the "FitzGerald contraction." The FitzGerald contraction introduced a factor that just neutralized the difference in time spent by the two light rays on their travel and therefore accounted for the absence of interference fringes. And yet the FitzGerald contraction had the appearance of a "gimmick factor." It worked, yes, but why should the contraction exist at all?

d) The Austrian physicist Ernst Mach went to the heart of the matter. He said there were no interference fringes because there was no ether wind because there was no ether. What could be simpler?

This was not a strange thing for Mach to have said. He was a rebel who insisted that only observable phenomena were rightly a matter for scientific inquiry, and that scientists should not set up models that were not themselves directly observable, and then believe in their actual existence. Mach even refused to accept atoms as anything more than a convenient fiction. Naturally, it was to be expected that he would be ready to scrap the ether the first chance he got.

How tempting that must have been! The ether was such a ridiculous and self-contradictory substance that some of the greatest nineteenth-century theoretical physicists had worn themselves out trying to explain it. Why not throw it away, as Mach irascibly suggested?

The trouble was—how would one then account for the fact that light could cross a vacuum? Everyone admitted that light consisted of waves, and the waves had to be waves of *something*. If the ether existed, light consisted of waves of ether. If the ether did not exist, then light consisted of waves of *what?*

Physics was hovering between the frying pan of ether

and the fire of complete chaos, and heaven only knows what would have happened if two German scientists, Max Karl Ernst Ludwig Planck in 1900, and Albert Einstein in 1905, had not come along to save the situation.

Save it, however, they did. The work of Planck and Einstein proved that light behaved as particles in some ways and that the ether therefore was not needed for light to travel through a vacuum. When this was done, the ether was no longer useful and it was dropped with a glad cry. The ether has never been required since. It does not exist now; in fact, it never existed. (Einstein's work also placed the FitzGerald contraction in the proper perspective.)

As a result, the Michelson-Morley light-that-failed was recognized as the most tremendously successful failure in the history of science, for it completely altered the physicists' view of the universe. In 1907 Michelson received the Nobel Prize in Physics for his optical labors generally, the first American to win one of the science prizes.

Eight
THE LIGHT FANTASTIC

When I was young, we children used to listen to something called "radio." It's a hard thing to describe to the modern population, but if you imagine a television set with the picture tube permanently out of order, you've got the essentials.

On the radio set there was a dial you could turn in order to tune in the various stations and the dial had markings numbered from 55 to 160. As far as I know, nobody I knew had any idea what those numbers meant—or cared.

A particular radio station might describe itself as possessing "880 kilocycles," and eventually I deduced that the numbers on the radio dial referred to tens of kilocycles, but again I never bumped into anyone (when I was young) who knew or cared what a kilocycle was.

In fact, as I look back upon it now, I don't think I knew or cared myself. I could dial any radio station I wanted with quick sureness and I had the radio schedule memorized. What more could I want?

And yet, if you consider the dial of a radio set, and proceed by free association, you can end up some pretty amazing matters, as I shall try to show you.

I'll begin with waves.

The most important waves in the universe are set up by oscillating electric charges. Since all electric charges have associated magnetic fields, the radiating waves produced in this fashion are called "electromagnetic." Electromagnetic waves radiate outward from the point of origin, moving at the velocity of light—which is not surprising, since light is itself an electromagnetic radiation.

Each oscillation of the electric charge back and forth gives rise to a single wave and from this fact we can calculate the length of the wave to which it gives rise. The length of a wave is called, with commendable simplicity, the "wavelength," and it is usually symbolized by the Greek letter *lambda* (λ).

Now suppose the electric charge oscillates once per second. By the time the end of the wave is formed at the completion of the oscillation, the beginning of the wave has been speeding out through space at the velocity of light one full second. The velocity of light in a vacuum is 186,200 miles per second or, in the metric system, which I shall use exclusively in this article, 300,000 kilometers per second. If, therefore, it takes a second to form the wave, the beginning of the wave is 300,000 kilometers ahead of the end of the wave and the wavelength is 300,000 kilometers.

Suppose the electric charge oscillates twice per second. Then in one second two waves are formed. Together they stretch out over 300,000 kilometers and each wave is 150,000 kilometers long; 150,000 kilometers is therefore the wavelength.

If the electric charge oscillates ten times per second, each wave is 30,000 kilometers long. If it oscillates fifty times per second, the wavelength is 6000 kilometers, and so on.

The number of oscillations per second can be called the "frequency," and this is usually symbolized by the Greek letter *nu* (ν).

As you see, what I have been doing in order to work out the wavelength of electromagnetic radiation is to divide the velocity of light (usually represented by c) by the frequency of the radiation. Put this in equation form and you have:

$$\lambda = \frac{c}{\nu}$$

If you know the wavelength and want to find the frequency, you need only solve for ν in the equation above, and you have:

$$\nu = \frac{c}{\lambda}$$

Thus, if the wavelength is 15 kilometers, then the frequency is $300{,}000/15$, or 20,000 oscillations per second.

A frequency of one oscillation per second can be described as one kilocycle (the prefix *kilo-*, being used in the metric system to represent "one thousand"). If, then, radio station WNBC in New York is located at 660 kilocycles (or at 66 on the dial), then that means the wave it puts out has a frequency of 660,000 oscillations per second. The wavelength of those waves is $^{300,000}/_{660,000}$ or 0.455 kilometers. This is equivalent to 455 meters.

In the same way, we can calculate the wavelengths of the waves put out by some other New York radio stations:

	Kilocycles	*Wavelength (meters)*
WOR	710	425
WABC	770	390
WNYC	830	360
WCBS	880	340
WNEW	1130	265
WQXR	1560	190

Notice that the wavelength gets shorter as the kilocycles increase; which is why, if we go up high enough on the dial, we end up with "short-wave radio." One way of expressing this relationship is to say that frequency and wavelength are inversely proportional to each other; as one goes up, the other goes down.

An electromagnetic radiation can have any wavelength, as far as we know, since a charged particle can oscillate at any frequency. There is no upper limit to the wavelength, certainly, for the oscillation can be slowed down to zero, in which case the wavelength approaches the infinite.

On the other hand, electric charges can be made to oscillate millions of times per second by man. Atoms can (in effect) oscillate trillions of times a second. Electrons can oscillate quadrillions and even quintillions of times per second. Nuclear particles can oscillate sextillions and even septillions of times per second. Wavelengths can get shorter and shorter, with no lower limit in theory.

The properties of electromagnetic radiation vary with frequency. For one thing, the radiation is put out in discrete little bundles called "quanta" and the energy content of one quantum of a particular radiation is in direct proportion to its frequency. As frequency goes up (and wavelength down) the radiation becomes more energetic and can interact more thoroughly with matter.

Short-wave radiation may knock electrons out of metals where longer-wave, less energetic radiation will not, and this is known as the photoelectric effect. (Einstein explained the rationale behind the photoelectric effect in 1905, the same year in which he first advanced his theory of relativity; and when he got his Nobel Prize in 1921, it was for his explaining the photoelectric effect, *not* for relativity.)

Again, short-wave radiation will bring about certain chemical changes where long-wave radiation will not, which is why you can develop ordinary photographic film under a red light. The red-light radiations are too low in energy to affect the negative.

Certain ranges of radiation are energetic enough to affect the retina of the eye and give us the sensation we call light. Radiation less energetic cannot be seen, but the energy can be absorbed by the skin and felt as heat. Radiation more energetic cannot be seen either, but can damage the retina and burn the skin.

It is convenient for physicists to divide the entire range of electromagnetic radiation (the "electromagnetic spectrum") into arbitrary regions. Here they are in the order of increasing frequency and energy and, therefore, of decreasing wavelength.

1) *Micropulsations.* These have frequencies of less than 1 cycle and, therefore, wavelengths of more than 300,000 kilometers. Such radiation has been detected with frequencies of as little as 0.01 cycle. This means that one oscillation takes 100 seconds and the wavelength is 30,000,000 kilometers, or three fourths of the way from here to Venus at its closest, which isn't bad for one wave.

2) *Radio waves.* In its broadest sense, these would include everything with frequencies from 1 cycle to 1 billion (10^9) cycles, and with wavelengths from 300,000 kilometers down to 30 centimeters. Actually, long-wave radio makes use of frequencies from 550,000 cycles to 1,600,000 cycles and wavelengths from 550 meters down to 185 meters. Short-wave radio uses wavelengths in the 30-meter range, and television in the 3-meter range.

3) *Microwaves.* The frequencies are from 1 billion (10^9) to 100 billion (10^{11}) cycles and the wavelengths are from 30 centimeters to 0.3 centimeters. The radiation detected by

radio telescopes is in this range and the radiation of the neutral hydrogen atom (the famous "song of hydrogen") has a wavelength of 21 centimeters. Radar also makes use of this range.

4) *Infrared rays.* The frequencies are from 100 billion (10^{11}) cycles to nearly a quadrillion (10^{14} plus) cycles and the wavelengths run from 0.3 centimeters to 0.000076 centimeters. Infrared wavelengths are usually measured in micrometers, one micrometer being a ten-thousandth of a centimeter, so the infrared wavelength range can be said to extend from 3000 micrometers down to 0.76 micrometers.

5) *Visible light rays.* These include a short stretch of frequencies just under the quadrillion mark (10^{15} minus), with wavelengths from 0.76 micrometers to 0.38 micrometers. Light wavelengths are usually measured in angstrom units, one angstrom unit being equal to a ten-thousandth of a micrometer. Thus, the wavelengths of visible light range from 7600 angstrom units down to 3800 angstrom units.

6) *Ultraviolet rays.* These include frequencies from a quadrillion (10^{15}) cycles up to nearly a hundred quadrillion (10^{17} minus) cycles, and the wavelengths run from 3800 angstrom units down to about 100 angstrom units.

7) *X-rays.* These include frequencies from nearly a hundred quadrillion (10^{17} minus) up to a hundred quintillion (10^{20}) cycles, with wavelengths ranging from 100 angstrom units down to 0.1 angstrom units.

8) *Gamma rays.* These make up the frequencies that are more than a hundred quintillion (10^{20}) cycles and wavelengths less than 0.1 angstrom units.

Actually, the dividing lines are anything but sharp, and X-rays and gamma rays, in particular, overlap generously. People speak of a particular frequency as being an X-ray if it is created in an X-ray tube and as a gamma ray if it is produced by a nuclear reaction. You can have soft gamma rays with wavelengths some three hundred times as long as the hardest X-rays. However, a particular wavelength has a particular energy and a particular set of properties regardless of what you call it: X-ray, gamma ray, or herring. By setting the boundary between X-rays and gamma rays at a frequency of a hundred quintillion cycles, I merely cut the overlap in half and am perfectly willing to admit the boundary is arbitrary.

Now this is a bewildering array of frequencies and wavelengths and I wouldn't be me if I didn't look for an easier way of presenting it. The easier way is drawn from usage in connection with sound waves. Sound waves are not electromagnetic in nature, but they, too, have wavelengths and frequencies.

We detect differences in the frequency of sound waves, at least in the audible range, by differences in the pitch we hear. It is conventional in our culture to write music using a series of notes with fixed frequencies. I will begin with the note on the piano which is called "middle C" and give its frequency and that of successive notes as we proceed toward the right on the keyboard, considering white keys only:

do — 264	do — 528	do — 1056
re — 267	re — 594	
mi — 330	mi — 660	
fa — 352	fa — 704	
sol — 396	sol — 792	
la — 449	la — 880	
ti — 495	ti — 990	

Notice that the frequency of each "do" is just twice the frequency of the preceding one. In fact, starting anywhere on the keyboard, one can progress through seven notes of increasing frequency and end with an eighth note of just twice the frequency of the first. Such a stretch is called an "octave," from the Latin word for "eight."

Applying this to any wave form in general, one can speak of an octave as applying to any continuous region stretching from frequency of x to one of $2x$. Since wavelength is inversely proportional to frequency, every time a frequency is doubled, a wavelength is halved. Every region stretching from a wavelength of y to one of $y/2$ is also an octave, therefore.

So we can break up the electromagnetic spectrum into octaves. As an example, the longest wavelengths of visible light are 7600 angstrom units, while the shortest are 3800 angstrom units. The shortest wavelengths are just half the longest and so the range covered by visible light is equal to one octave of the electromagnetic spectrum.

Since there is no upper or lower limit to the frequencies of the electromagnetic spectrum, the number of octaves is, theoretically, infinite. However, suppose we consider a wavelength of 30,000,000 kilometers as the practical maximum, since this is the longest micropulsation detected, and a wavelength of 0.0001 angstrom units as the practical minimum, since beyond that lie the energy ranges associated with cosmic rays, which are particulate rather than electromagnetic in nature.

The number of times you must halve 30,000,000 kilometers to reach 0.0001 angstrom units is 81. (Try it and see, and remember that 1 kilometer equals 10,000,000,000,000 angstrom units.) The portion of the electromagnetic spectrum I have marked off, therefore, is eighty-one octaves long, and of that length, we see exactly one octave with our eyes.

Now let's measure off the confusing divisions of the electromagnetic spectrum in octaves and the picture will be much simpler:

	Octaves
micropulsations	6½
radio waves	30
microwaves	6½
infrared rays	12
visible light rays	1
ultraviolet rays	5
X-rays	10
gamma rays	10
total	81

As you see, two thirds of the octaves are longer-wave and, therefore, less energetic than light. In fact, the radio-wave region at its broadest takes up a third of the octaves of the spectrum. Actually, though, only about twelve octaves altogether are used for radio and television communications.

Still that makes up about 15 per cent of the total number of octaves and as our needs for communication increase with the developing space age, how much room for expansion can there be?

The answer is: Plenty!

To see why that is, let's consider this matter of octaves further. In the realm of sound, the ear finds all octaves equal. In each one, there is room for seven different notes (plus sharps and flats, of course) before the next octave begins.

This is not so, however, as far as communication by

electromagnetic waves is concerned. As one goes up the electromagnetic spectrum in the direction of increasing frequency, each octave has more room than the one before.

Each television channel emits a carrier wave which it modifies, these modifications being converted into sight and sound at the receiving television set. In order for two channels not to interfere with each other, they must have frequencies that are not too close. They can't be anywhere near as closely spaced as the radio stations with which I began this article, for instance. The width of a standard television channel is 4,000,000 cycles (or 4 megacycles, a megacycle being equal to a million cycles).

The television channels fall at the short-wave end of the radio-wave region, in the range of frequencies 100,000,000 cycles (100 megacycles) and wavelengths of about 3 meters.

Consider an octave in this region of frequencies; say a stretch of the spectrum from a frequency of 80 megacycles to one of 160 megacycles. This covers a width of 80 megacycles, and if television channels are spaced at 4 megacycle intervals, there is room for twenty channels.

In the next octave up, from 160 to 320 megacycles, there is room for forty channels. In the one after that, from 320 to 640 megacycles, there is room for eighty channels.

The number of television channels per octave of electromagnetic radiation doubles for each successive octave as one moves up the scale in the direction of increasing frequency. In fact, each octave of electromagnetic radiation contains about as much room for television channels as do all the preceding octaves put together.

What about visible light, then? There is only one octave of visible light, but it is roughly twenty-two octaves higher than the one used for television. There is thus 2^{22} times as much room for television channels in the octave of light as in the octave ordinarily used for television. The figure 2^{22} represents the product of twenty-two 2's, and that comes to over four million. (Multiply them out for yourself; don't take my word for it.)

In other words, for every channel available in the usual television portion of the electromagnetic spectrum, there would be some four million channels available in the visible light portion.

We can break this down in more detail. The visible spectrum contains a number of colors that fade one into the other as you go up or down the scale. Actually, the eye can distinguish among a great number of shades and there are no sharp boundaries. Nevertheless, it is customary to divide the visible spectrum into six colors, which, in order of increasing frequency, are red, orange, yellow, green, blue, and violet. And each color is considered as stretching over a certain range of frequency. The situation might be presented thus:

	Wavelength Range (angstrom units)	*Frequency Range (megacycles)*
red	7600 to 6300	400,000,000 to 475,000,000
orange	6300 to 5900	475,000,000 to 510,000,000
yellow	5900 to 5600	510,000,000 to 540,000,000
green	5600 to 4900	450,000,000 to 615,000,000
blue	4900 to 4500	615,000,000 to 670,000,000
violet	4500 to 3800	670,000,000 to 800,000,000

Remembering that the width of a standard television channel is only 4 megacycles, then we can set up the following table:

	Width of Frequency Band (megacycles)	*Number of Television Channels Possible*
red	75,000,000	19,000,000
orange	35,000,000	9,000,000
yellow	30,000,000	7,000,000
green	75,000,000	19,000,000
blue	55,000,000	14,000,000
violet	130,000,000	32,000,000
	total	100,000,000

Well, then, why not use light waves as carriers for television broadcasts?

Until two years ago* this was a suggestion that could have only theoretical interest. The carrier waves set up for ordinary radio-television communication can be produced in perfect phase. They form an orderly succession of waves that can be neatly modified in any fashion.

Light waves, however, cannot be set up so neatly in phase; at least they couldn't until the 1960s. It is quite impractical to try to oscillate an electric-circuit five hundred

* This article first appeared in August 1962.

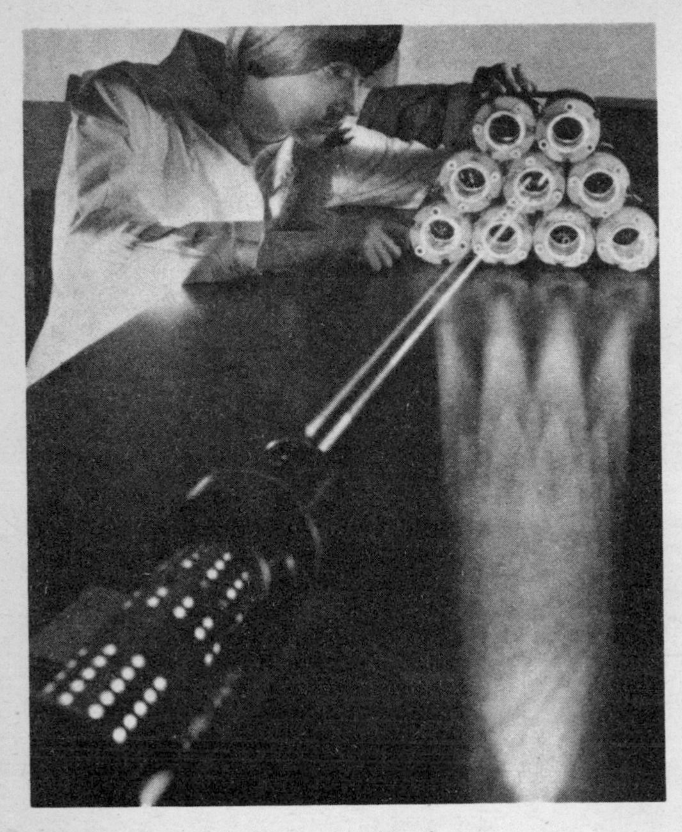

Photo by John T. Hill. Courtesy: Perkin–Elmer.

THE LASER

This article was written not long after the invention of the laser—an example of which, one that makes use of a gas as the "lasing" substance, is shown here. At that time I saw only its use as a communications device and that, largely, between ships in space. With more than a decade of hindsight, I can see that much more has turned up.

(1) The light waves produced by a laser are much more regular than any other kind of light waves in the world. The laser light, as far as information is concerned, is a blank page—a given direction, a given wavelength, and that's all. Ordinary light (with a variety of wavelengths and directions) is full of information and if it intersects with certain light-sensitive chemicals on a flat surface, the result is a photograph. The photograph is purely two-dimensional and gives only a tiny bit of the total information of the light.

If the light waves intersect with laser light, interference fringes can be produced and those can be recorded on a photographic negative. The interference fringes, representing the intersection of light waves with light waves, contain all the information, and if light is sent through the negative the interference is re-created and the light passes through to form a three-dimensional "holographic image."

(2) Heavy hydrogen, or deuterium, could be made to fuse in a controlled way, giving humanity a permanent source of cheap energy, if only it can be heated high enough quickly enough. Laser light can be focused so tightly that enough energy can be poured into a tiny spot to heat it to millions of degrees in a very tiny fraction of a second. Several laser beams concentrated on one little bubble of hydrogen under proper, rather complicated, conditions could initiate fusion. The laser, therefore, may serve as the trigger for the solution of the energy crisis.

trillion times a second, which is what would be required to send out a beam of visible light. The electrons within an atom must be relied upon for such an oscillation. Heat is poured into them and it is liberated as electromagnetic radiation, much of which (because of their natural rate of oscillation) is in the form of visible light. In other words, you can make light by starting a fire.

The only trouble is that the various heated atoms give off radiation, each in its own good time, and the wavelength is not fixed but can be varied over a wide range, and the quantum is fired out in any direction. Thus, the emitted light waves are so much out of phase that most of their energy is canceled and converted into heat; they spread out widely in every direction and cover a broad range of the spectrum. In short, the light produced is good enough to see by, but not good enough to serve as a carrier wave for TV.

However, in 1960 instruments were devised into which energy could be pumped and then, when a sparking bit of light was allowed to enter, all the energy was converted into light of the same wavelength, and all in phase. The device could be so constructed that all the light would emerge in the same direction, too.

The beam of intense light that is produced by such a device would stick together (it would be "coherent") and it would possess an extremely narrow band of wavelengths (it would be "monochromatic"). The process by which a bit of light sparks the conversion of energy into a lot of light is called "*l*ight *a*mplification by *s*timulated *e*mission of *r*adiation," and by taking the indicated initials, the instrument was named a "laser." (In case you are interested, a word constructed out of the initials of a phrase is called an "acronym.")

Of course, even so, the use of light as television carrier waves presents difficulties. In the range of the electromagnetic spectrum currently used for television, the radiation can penetrate buildings and go through ordinary obstacles. Visible light can't do this. You would need a clear and unobstructed view of the TV station before you could receive a program.

It is possible, however, that light might be sent through plastic pipes, from which leads could reach each television set in the area. (Does that mean the streets all get dug up, or will the pipes run along telephone poles, or what?)

On the other hand, television by laser would be ideal out in space, where ship could reach ship or space station through the uninterrupted reaches of vacuum, and each ship could have a television channel reserved all for itself. It would be a long time before we had more than a hundred million ships out in space, so there would be no crowding. Then, even if we did run out of room in the visible region, the ultra-violet portion of the spectrum would give room for about six billion more channels.

Of course, there is something else—

These days, when I watch television here at home, I have my choice of four channels that I can get with reasonable clearness and audibility. Even with only four channels at their disposal, however, the television moguls can supply me with a tremendous quantity of rubbish.

Imagine what the keen minds of our entertainment industry could do if they realized they had a hundred million channels into which they could funnel new and undreamed-of varieties of trash.

Maybe we ought to stop right now!†

† Thirteen years later we are not measurably closer to this fearful fate although we now have working communication-satellites which, in conjunction with laser beams, may eventually utterly revolutionize our way of life. That's probably some time off, however. At the present moment the most exciting prospect for lasers is that they offer us a possible way of initiating a controlled fusion reaction for a new and immensely large energy-source.

Nine
C FOR CELERITAS

If ever an equation has come into its own it is Einstein's $e = mc^2$. Everyone can rattle it off now, from the highest to the lowest; from the rarefied intellectual height of the science-fiction reader, through nuclear physicists, college students, newspaper reporters, housewives, busboys, all the way down to congressmen.

Rattling it off is not, of course, the same as understanding it; any more than a quick paternoster (from which, incidentally, the word "patter" is derived) is necessarily evidence of deep religious devotion.

So let's take a look at the equation. Each letter is the initial of a word representing the concept it stands for. Thus, e is the initial letter of "energy" and m of "mass." As for c, that is the speed of light in a vacuum, and if you ask why c, the answer is that it is the initial letter of *celeritas*, the Latin word meaning "speed."

This is not all, however. For any equation to have meaning in physics, there must be an understanding as to the units being used. It is meaningless to speak of a mass of 2.3, for instance. It is necessary to say 2.3 grams or 2.3 pounds or 2.3 tons; 2.3 alone is worthless.

Theoretically, one can choose whatever units are most convenient, but as a matter of convention, one system used in physics is to start with "grams" for mass, "centimeters" for distance, and "seconds" for time; and to build up, as far as possible, other units out of appropriate combinations of these three fundamental ones.

Therefore, the m in Einstein's equation is expressed in grams, abbreviated gm. The c represents a speed—that is, a distance traveled in a certain time. Using the fundamental units, this means the number of centimeters traveled in a certain number of seconds. The units of c are therefore centimeters per second, or cm/sec.

(Notice that the word "per" is represented by a fraction line. The reason for this is that to get a speed represented in lowest terms, that is, the number of centimeters traveled in *one* second, you must divide the number of centimeters traveled by the number of seconds of traveling. If you travel 24 centimeters in 8 seconds, your speed is 24 centimeters ÷ 8 seconds, or 3 cm/sec.)

But, to get back to our subject, c occurs as its square in the equation. If you multiply c by c, you get c^2. It is, however, insufficient to multiply the numerical value of c by itself. You must also multiply the unit of c by itself.

A common example of this is in connection with measurements of area. If you have a tract of land that is 60 feet by 60 feet, the area is not 60 × 60, or 3600 feet. It is 60 feet × 60 feet, or 3600 square feet.

Similarly, in dealing with c^2, you must multiply cm/sec by cm/sec and end with the units cm^2/sec^2 (which can be read as centimeters squared per seconds squared).

The next question is: What is the unit to be used for e? Einstein's equation itself will tell us, if we remember to treat units as we treat any other algebraic symbols. Since $e = mc^2$, that means the unit of e can be obtained by multiplying the unit of m by the unit of c^2. Since the unit of m is gm and that of c^2 is cm^2/sec^2, the unit of e is gm × cm^2/sec^2. In algebra we represent $a \times b$ as ab; consequently, we can run the multiplication sign out of the unit of e and make it simply gm cm^2/sec^2 (which is read "gram centimeter squared per second squared).

As it happens, this is fine, because long before Einstein worked out his equation it had been decided that the unit of energy on the gram-centimeter-second basis had to be gm cm^2/sec^2. I'll explain why this should be.

The unit of speed is, as I have said, cm/sec, but what happens when an object changes speed? Suppose that at a given instant, an object is traveling at 1 cm/sec, while a second later it is traveling at 2 cm/sec; and another second later it is traveling at 3 cm/sec. It is, in other words, "accelerating" (also from the Latin word *celeritas*).

In the case I've just cited, the acceleration is 1 centimeter per second every second, since each successive second it is going 1 centimeter per second faster. You might say that the acceleration is 1 cm/sec per second. Since we are letting the word "per" be represented by a fraction mark, this may be represented as 1 cm/sec/sec.

As I said before, we can treat the units by the same

manipulations used for algebraic symbols. An expression like $a/b/b$ is equivalent to $a/b \div b$, which is in turn equivalent to $a/b \times 1/b$, which is in turn equivalent to a/b^2. By the same reasoning, 1 cm/sec/sec is equivalent to 1 cm/sec^2 and it is cm/sec^2 that is therefore the unit of acceleration.

A "force" is defined, in Newtonian physics, as something that will bring about an acceleration. By Newton's First Law of Motion any object in motion, left to itself, will travel at constant speed in a constant direction forever. A speed in a particular direction is referred to as a "velocity," so we might, more simply, say that an object in motion, left to itself, will travel at constant velocity forever. This velocity may well be zero, so that Newton's First Law also says that an object at rest, left to itself, will remain at rest forever.

As soon as a force, which may be gravitational, electromagnetic, mechanical, or anything, is applied, however, the velocity is changed. This means that its speed of travel or its direction of travel or both is changed.

The quantity of force applied to an object is measured by the amount of acceleration induced, and also by the mass of the object, since the force applied to a massive object produces less acceleration than the same force applied to a light object. (If you want to check this for yourself, kick a beach ball with all your might and watch it accelerate from rest to a good speed in a very short time. Next kick a cannon ball with all your might and observe—while hopping in agony—what an unimpressive acceleration you have imparted to it.)

To express this observed fact, one uses the expression: "Force equals mass times acceleration" or, to abbreviate, $f = ma$. Since the unit of mass is gm and the unit of acceleration is cm/sec^2, the unit of force is the product of the two or gm cm/sec^2.

Physicists grow tired of muttering "gram centimeter per second squared" every other minute, so they invented a single syllable to represent that phrase. The syllable is *dyne*, from the Greek *dynamics* meaning "power."

The multisyllabic expression and the monosyllable are equivalent: 1 dyne = 1 gm cm/sec^2. Dyne is just a breath-saver and can be defined as follows: A dyne is that amount of force which will impose upon a mass of one gram an acceleration of one centimeter per second squared.

See?

Next, there arises the problem of "work." Work as defined by the physicist is not what I do when I sit at the typewriter to write a chapter and slave my head to the very bone. To the physicist, "work" is simply the motion of a body against a resting force. To lift an object against the force of gravity is work; to pull a bar of iron away against the pull of a magnet is work; to drive a nail into the wood against the resistance of friction is work; and so on.

The amount of work depends on the size of the resisting force and the distance moved against it. This can be expressed by saying: "Work equals force times distance," or, by abbreviation, $w = fd$.

The unit of distance is cm and the unit of force is dyne. Consequently, the unit of work is dyne cm. Again, physicists invented a monosyllable to express "dyne centimeters," and the new monosyllable is the ugly sound *erg*, from the Greek *ergon* meaning "work."

An erg is defined as the unit of work, and 1 erg is the amount of work performed by moving an object one cm against the resisting force of one dyne.

Lest you forget that this is all based on the gram-centimeter-second system, bring to mind the fact that a dyne is equivalent to a gm cm/sec^2. This means that the unit of work is cm times gm cm/sec^2 (distance times force), and this works out to gm cm^2/sec^2. In other words, 1 erg is the work done by imposing upon a mass of 1 gm an acceleration of 1 cm/sec^2 over a distance of 1 cm.

It was discovered a little over a century ago that work and energy are interconvertible, so that the units for one will serve as the units for the other. Consequently, the erg is also the unit of energy on the gram-centimeter-second basis.

Now shall we get back to Einstein's equation? There the units of e worked out to gm cm^2/sec^2, and that is equivalent to ergs. Those are the units we expect for energy, and it's no coincidence. If the equation had worked out to give any other units for energy, Einstein would have sharpened his pencil and started over again, knowing he had made a mistake.

Now we are ready to put numerical values into Einstein's equation. As far as m is concerned, we can suit ourselves and choose any convenient numerical quantity, the simplest choice being 1 gm.

In the case of c we have no option. The speed of light in a vacuum has a certain value and no other. In the units we have decided on, the best figure we have today is 29,979,000,000 cm/sec.* We wouldn't be far wrong in rounding this off to 30,000,000,000 cm/sec (a speed at which light can cover thirty billion centimeters—or three-quarters of the distance to the moon—in one second). Exponentially we can express this as 3×10^{10} cm/sec.

We have to square this to get the value of c^2, remembering to square both the number and the unit, and we end with 900,000,000,000,000,000,000 cm²/sec² or 9×10^{20} cm^2/sec^2. The expression mc^2 (which is equal to e in Einstein's equation) thus becomes: 1 gm $\times 9 \times 10^{20}$ cm^2/sec^2, which works out to 9×10^{20} gm cm^2/sec^2 or, if you prefer, 9×10^{20} ergs.

In other words, if 1 gram of matter were completely converted to energy, you would find yourself possessed of nine hundred quintillion ergs. And, on the other hand, if you wished to create 1 gram of matter out of pure energy (and could manage it with perfect efficiency), you would have to assure yourself, first, of a supply of nine hundred quintillion ergs.

This sounds impressive. Nine hundred quintillion ergs, wow!

But then, if you are cautious, you might stop and think: An erg is an unfamiliar unit. How large is it anyway?

After all, in Al Capp's Lower Slobbovia, the sum of a billion slobniks sounds like a lot until you find that the rate of exchange is ten billion slobniks to the dollar.

So— How large is an erg?

Well, it isn't large. As a matter of fact, it is quite a small unit. It is forced on physicists by the logic of the gram-centimeter-second system of units, but it ends in being so small a unit as to be scarcely useful. For instance, consider the task of lifting a pound weight one foot against gravity. That's not difficult and not much energy is expended. You could probably lift a hundred pounds one foot without completely incapacitating yourself. A professional strong man could do the same for a thousand pounds.

Nevertheless, the energy expended in lifting *one* pound

* This chapter first appeared in November 1959. Now, fifteen years later, the best figure is 29,979,245,620 cm/sec.

one foot is equal to 13,558,200 ergs. Obviously, if any trifling bit of work is going to involve ergs in the tens of millions, we need other and larger units to keep the numerical values conveniently low.

For instance, there is an energy unit called a *joule*, which is equal to 10,000,000 ergs.

This unit is derived from the name of the British physicist James Prescott Joule, who inherited wealth and a brewery but spent his time in research. From 1840 to 1849 he ran a series of meticulous experiments which demonstrated conclusively the quantitative interconversion of heat and work and brought physics an understanding of the law of conservation of energy. However, it was the German scientist Hermann Ludwig Ferdinand von Helmholtz who first put the law into actual words in a paper presented in 1847, so that he consequently gets formal credit for the discovery.

(The word "joule," by the way, is most commonly pronounced "jowl," although Joule himself probably pronounced his name "jool." In any case, I have heard overprecise people pronounce the word "zhool" under the impression that it is a French word, which it isn't. These are the same people who pronounce "centigrade" and "centrifuge" with a strong nasal twang as "sontigrade" and "sontrifuge," under the impression that these, too, are French words. Actually, they are from Latin and no pseudo-French pronunciation is required. There is some justification for pronouncing "centimeter" as "sontimeter," since that is a French word to begin with, but in that case one should either stick to English or go French all the way and pronounce it "sontimettre," with a light accent on the third syllable.)

Anyway, notice the usefulness of the joule in everyday affairs. Lifting a pound mass a distance of one foot against gravity requires energy to the amount, roughly, of 1.36 joules—a nice, convenient figure.

Meanwhile, physicists who were studying heat had invented a unit that would be convenient for their purposes. This was the "calorie" (from the Latin word *calor* meaning "heat"). It can be abbreviated as cal. A calorie is the amount of heat required to raise the temperature of 1 gram of water from 14.5° C. to 15.5° C. (The amount of heat necessary to raise a gram of water one Celsius degree

The Granger Collection

HERMANN LUDWIG FERDINAND VON HELMHOLTZ

Helmholtz, the son of a schoolteacher, was born at Potsdam, Germany, on August 31, 1821. He had a sickly childhood and suffered all his life from migraine headaches and fainting spells.

He obtained his medical degree in 1842 and was particularly interested in the sense organs. In 1851, he invented the ophthalmoscope, an instrument which could peer into the eye's interior. He worked out the theory of three-color vision, advanced a theory as to the working of the inner ear, and pointed out the importance of overtones in sound.

Helmholtz was the first to measure the speed of the nerve impulse. His teacher in biology had been fond of presenting this as an example of something science could never accomplish because the impulse moved so quickly over so short a path. Helmholtz managed to make the measurement, however, in 1852.

He is best known for his contributions to physics and, in particular, for his support of the law of conservation of energy (something to which he was led by his studies of muscle action).

Considering the law of conservation of energy, Helmholtz was the first to take up the question of the source of the Sun's energy. He finally decided that the only possible source was gravitational energy and maintained that the Sun must be steadily contracting as it shines. In the past 25 million years, he suggested, it had contracted, in this way, from a size large enough to fill the Earth's orbit. This meant that the Earth was only 25 million years old—not enough to account for geological changes and biological evolution. The dilemma was not solved till the discovery of nuclear energy.

Helmholtz died near Berlin, on September 8, 1894, after suffering a concussion as a result of falling during a fainting spell.

varies slightly for different temperatures, which is why one must carefully specify the 14.5 to 15.5 business.)

Once it was demonstrated that all other forms of energy and all forms of work can be quantitatively converted to heat, it could be seen that any unit that was suitable for heat would be suitable for any other kind of energy or work.

By actual measurement it was found (by Joule) that 4.185 joules of energy or work could be converted into precisely 1 calorie of heat. Therefore, we can say that 1 cal equals 4.185 joules 41,850,000 ergs.

Although the calorie, as defined above, is suitable for physicists, it is a little too small for chemists. Chemical reactions usually release or absorb heat in quantities that, under the conventions used for chemical calculations, result in numbers that are too large. For instance, 1 gram of carbohydrate burned to carbon dioxide and water (either in a furnace or the human body, it doesn't matter) liberates roughly 4000 calories. A gram of fat would, on burning, liberate roughly 9000 calories. Then again, a human being, doing the kind of work I do, would use up about 2,500,000 calories per day.

The figures would be more convenient if a larger unit were used, and for that purpose a larger calorie was invented, onc that would represent the amount of heat required to raise the temperature of 1000 grams (1 kilogram) of water from 14.5° C. to 15.5° C. You see, I suppose, that this larger calorie is a thousand times as great as the smaller one. However, because both units are called "calorie," no end of confusion has resulted.

Sometimes the two have been distinguished as "small calorie" and "large calorie"; or "gram-calorie" and "kilogram-calorie"; or even "calorie" and "Calorie." (The last alternative is a particularly stupid one, since in speech—and scientists must occasionally speak—there is no way of distinguishing a *C* and a *c* by pronunciation alone.)

My idea of the most sensible way of handling the matter is this: In the metric system, a kilogram equals 1000 grams; a kilometer equals 1000 meters, and so on. Let's call the large calorie a kilocalorie (abbreviated kcal) and set it equal to 1000 calories.

In summary, then, we can say that 1 kcal equals 1000 cal or 4185 joules or 41,850,000,000 ergs.

Another type of energy unit arose in a roundabout way, via the concept of "power." Power is the rate at which work is done. A machine might lift a ton of mass one foot against gravity in one minute or in one hour. In each case the energy consumed in the process is the same, but it takes a more powerful heave to lift that ton in one minute than in one hour.

To raise one pound of mass one foot against gravity takes one *foot-pound* (abbreviated 1 ft-lb) of energy. To expand that energy in one second is to deliver 1 foot-pound per second (1 ft-lb/sec) and the ft-lb/sec is therefore a permissible unit of power.

The first man to make a serious effort to measure power accurately was James Watt (1736–1819). He compared the power of the steam engine he had devised with the power delivered by a horse, thus measuring his machine's rate of delivering energy in *horsepower* (or hp). In doing so, he first measured the power of a horse in ft-lb/sec and decided that 1 hp equals 550 ft-lb/sec, a conversion figure which is now standard and official.

The use of foot-pounds per second and horsepower is perfectly legitimate and, in fact, automobile and airplane engines have their power rated in horsepower. The trouble with these units, however, is that they don't tie in easily with the gram-centimeter-second system. A foot-pound is 1.355282 joules and a horsepower is 10.688 kilocalories per minute. These are inconvenient numbers to deal with.

The ideal gram-centimeter-second unit of power would be ergs per second (erg/sec). However, since the erg is such a small unit, it is more convenient to deal with joules per second (joule/sec). And since 1 joule is equal to 10,000,000 ergs, 1 joule/sec equals 10,000,000 erg/sec, or 10,000,000 gm cm^2/sec^3.

Now we need a monosyllable to express the unit joule/sec, and what better monosyllable than the monosyllablic name of the gentleman who first tried to measure power. So 1 joule/sec was set equal to 1 *watt*. The watt may be defined as representing the delivery of 1 joule of energy per second.

Now if power is multiplied by time, you are back to energy. For instance, if 1 watt is multiplied by 1 second, you have 1 *watt-sec*. Since 1 watt equals 1 joule/sec, 1 watt-sec equals 1 (joule/sec) × sec, or 1 joule sec/sec. The secs cancel as you would expect in the ordinary algebraic manipulation to which units can be subjected, and you end with the statement that 1 watt-sec is equal to 1 joule and is, therefore, a unit of energy.

A larger unit of energy of this sort is the *kilowatt-hour* (or kw-hr). A kilowatt is equal to 1000 watts and an hour is equal to 3600 seconds. Therefore a kw-hr is equal to 1000 × 3600 watt-sec, or to 3,600,000 joules, or to 36,000,000,000,000 ergs.

Furthermore, since there are 4185 joules in a kilocalorie (kcal), 1 kw-hr is equal to 860 kcal or to 860,000 cal.

A human being who is living on 2500 kcal/day is delivering (in the form of heat, eventually) about 104 kcal/hr, which is equal to 0.120 kw hr/hr or 120 watts. Next time you're at a crowded cocktail party (or a crowded subway

The Granger Collection

THE STEAM ENGINE

In a surprising number of ways, mankind's advances have been different from those of other living species only in degree, not in kind. Other animals communicate, though not as well as man. Other animals use tools, conduct wars, take slaves, herd other species, make use of the energy of wind and water.

No other species, however, in all the long history of life on Earth ever made use of fire. That single discovery was the beginning of Technological Man. It meant that the history of life had taken a major turning point; perhaps the most fateful one in all its history.

Yet for uncounted thousands of years after human beings had tamed fire, it was used only directly for its heat and light. It made the night safer and the winter endurable. It cooked food and made it more digestible, less parasite-ridden, and tastier. It smelted ore, baked pottery, formed glass, and so on.

It was not until nearly 1700, however, that attempts were made to use its energy to do the kind of work human and animal muscles had been doing; and it was not until nearly 1800 that a truly practical "steam engine" was devised. (The one shown here was built in 1876.) The steam engine was the first "prime mover" (something that converted inanimate energy into useful work) that made use of heat, and it marked the beginning of the Industrial Revolution.

Through most of the nineteenth century, the steam engine gradually took over the task of wind and current and pushed ships in the direction man willed; and the task of the horse, driving trains of cars over shining rails; and the task of human beings, running innumerable machines in factories. It even inspired the development of the science of thermodynamics and aided in the discovery of some of the basic laws of nature.

train or a crowded theater audience) on a hot evening in August, think of that as each additional person walks in. Each entrance is equivalent to turning on another one hundred twenty-watt electric bulb. It will make you feel a lot hotter and help you appreciate the new light of understanding that science brings.

But back to the subject. Now, you see, we have a variety of units into which we can translate the amount of energy resulting from the complete conversion of 1 gram of mass. That gram of mass will liberate:

900,000,000,000,000,000,000 ergs,
or 90,000,000,000,000 joules,
or 21,500,000,000,000 calories,
or 21,500,000,000 kilocalories,
or 25,000,000 kilowatt-hours.

Which brings us to the conclusion that although the erg is indeed a tiny unit, nine hundred quintillion of them still mount up most impressively. Convert a mere one gram of mass into energy and use it with perfect efficiency and you can keep a thousand-watt electric light bulb running for 25,000,000 hours, which is equivalent to 2850 years, or the time from the days of Homer to the present.

How's that for solving the fuel problem?

We could work it the other way around, too. We might ask: How much mass need we convert to produce 1 kilowatt-hour of energy?

Well, if 1 gram of mass produces 25,000,000 kilowatt-hours of energy, then 1 kilowatt-hour of energy is produced by 1/25,000,000 gram.

You can see that this sort of calculation is going to take us into small mass units indeed. Suppose we choose a unit smaller than the gram, say the *microgram*. This is equal to a millionth of a gram, i.e. 10^{-6} gram. We can then say that 1 kilowatt-hour of energy is produced by the conversion of 0.04 micrograms of mass.

Even the microgram is an inconveniently large unit of mass if we become interested in units of energy smaller than the kilowatt-hour. We could therefore speak of a *micro-microgram* (or, as it is now called, a *picogram*). This is a millionth of a millionth of a gram (10^{-12} gram) or a trillionth of a gram. Using that as a unit, we can say that:

1 kilowatt-hour	is	equivalent	to	40,000	picograms
1 kilocalorie	"	"	"	46.5	"
1 calorie	"	"	"	0.0465	"
1 joule	"	"	"	0.0195	"
1 erg	"	"	"	0.00000000195	"

To give you some idea of what this means, the mass of a typical human cell is about 1000 picograms. If, under conditions of dire emergency, the body possessed the ability to convert mass to energy, the conversion of the contents of 152 selected cells (which the body, with 50,000,000,000,000 cells or so, could well afford) would supply the body with 2500 kilocalories and keep it going for a full day.

The amount of mass which, upon conversion, yields 1 erg of energy (and the erg, after all, is the proper unit of energy in the gram-centimeter-second system) is an inconveniently small fraction even in terms of picograms.

We need units smaller still, so suppose we turn to the *picopicogram* (10^{-24} gram), which is a trillionth of a trillion of a gram, or a septillionth of a gram. Using the picopicogram, we find it takes the conversion of 1950 picopicograms of mass to produce an erg of energy.

And the significance? Well, a single hydrogen atom has a mass of about 1.66 picopicograms. A uranium-235 atom has a mass of about 400 picopicograms. Consequently, an erg of energy is produced by the total conversion of 1200 hydrogen atoms or by 5 uranium-235 atoms.

In ordinary fission, only 1/1000 of the mass is converted to energy so it takes 5000 fissioning uranium atoms to produce 1 erg of energy. In hydrogen fusion, 1/100 of the mass is converted to energy, so it takes 120,000 fusing hydrogen atoms to produce 1 erg of energy.

And with that, we can let $e = mc^2$ rest for the nonce.

Ten

THE ULTIMATE SPLIT OF THE SECOND

Occasionally, I get an idea for something new in science; not necessarily something important, of course, but new anyway. One of these ideas is what I will devote this chapter to.

The notion came to me some time ago, when the news broke that a subatomic particle called "xi-zero" (with "xi" pronounced "ksee," if you speak Greek, and "zigh" if you speak English) had been detected for the first time. Like other particles of its general nature, it is strangely stable, having a half-life of fully a ten-billionth (10^{-10}) of a second or so.

The last sentence may seem misprinted—you may think that I meant to write "unstable"—but no! A ten-billionth of a second can be a long time; it all depends on the scale of reference. Compared to a sextillionth (10^{-23}) of a second, a ten-billionth (10^{-10}) of a second is an aeon. The difference between those two intervals of time is as that between one day and thirty billion years.

You may grant this and yct feel an onset of dizziness. The world of split-seconds and split-split-split-seconds is a difficult one to visualize. It is easy to say "a sextillionth of a second" and just as easy to say "a ten-billionth of a second"; but no matter how easily we juggle the symbols representing such time intervals, it is impossible (or *seems* impossible) to visualize either.

My idea is intended to make split-seconds more visualizable, and I got it from the device used in a realm of measurement that is also grotesque and also outside the range of all common experience—that of astronomical distances.

There is nothing strange in saying, "Vega is a very nearby star. It's not very much more than a hundred fifty trillion (1.5×10^{14}) miles away."

Most of us who read science fiction are well-used to the thought that a hundred fifty trillion miles is a very small distance on the cosmic scale. The bulk of the stars in our galaxy are something like two hundred quadrillion (2×10^{17}) miles away, and the nearest full-sized outside galaxy is more than ten quintillion (10^{19}) miles away.

Trillion, quadrillion and quintillion are all legitimate number-words, and there's no difficulty telling which is larger and by how much, if you simply want to manipulate symbols. Visualizing what they mean, however, is another thing.

So the trick is to make use of the speed of light to bring the numbers down to vest-pocket size. It doesn't change the actual distance any, but it's easier to make some sort of mental adjustment to the matter if all the zeroes of the "-illions" aren't getting in the way.

The velocity of light in a vacuum is 186,274 miles per second or, in the metric system, 299,779 kilometers per second.*

A "light-second," then, can be defined as that distance which light (in a vacuum) will travel in a second of time, and is equal to 186,274 miles or to 299,779 kilometers.

It is easy to build longer units in this system. A "light-minute" is equal to 60 light-seconds; a "light-hour" is equal to 60 light-minutes, and so forth, till you reach the very familiar "light-year," which is the distance which light (in a vacuum) will travel in a year. This distance is equal to 5,890,000,000,000 miles or 9,460,000,000,000 kilometers. If you are content with round numbers, you can consider a light-year equal to six trillion (6×10^{12}) miles or nine and one-half trillion (9.5×10^{12}) kilometers.

You can go on, if you please, to "light-centuries" and "light-millennia," but hardly anyone ever does. The light-year is the unit of preference for astronomic distances. (There is also the "parsec," which is equal to 3.26 light-years, or roughly twenty trillion miles, but that is a unit based on a different principle, and we need not worry about it here.)

Using light-years as the unit, we can say that Vega is 27 light-years from us, and that this is a small distance con-

* Remember that I've given the current best figure for the speed of light in a previous footnote.

sidering that the bulk of the stars of our galaxy are 35,000 light-years away and that the nearest full-sized outside galaxy is 2,100,000 light-years away. The difference between 27 and 35,000 and 2,100,000, given our range of experience, is easier to visualize than that between a hundred fifty trillion and two hundred quadrillion and ten quintillion, though the ratios in both cases are the same.

Furthermore, the use of the speed of light in defining units of distance has the virtue of simplifying certain connections between time and distance.

For instance, suppose an expedition on Ganymede is, at a certain time, 500,000,000 miles from earth. (The distance, naturally, varies with time, as both worlds move about in their orbits.) This distance can also be expressed as 44.8 light-minutes.

What is the advantage of the latter expression? For one thing, 44.8 is an easier number to say and handle than 500,000,000. For another, suppose our expedition is in radio communication with earth. A message sent from Ganymede to earth (or vice versa) will take 44.8 minutes to arrive. The use of light-units expresses distance *and* speed of communication at the same time.

(In fact, in a world in which interplanetary travel is a taken-for-granted fact, I wonder if the astronauts won't start measuring distance in "radio-minutes" rather than light-minutes. Same thing, of course, but more to the point unless we start using light beams for communication, as mentioned in Chapter 8).

Then, when and if interstellar travel comes to pass, making necessary the use of velocities at near the speed of light, another advantage will come about. If time dilatation exists and the experience of time is slowed at high velocities, a trip to Vega may seem to endure for only a month or for only a week. To the stay-at-homes on earth, however, who are experiencing "objective time" (the kind of time that is experienced at low velocities—strictly speaking, at zero velocity), the trip to Vega, 27 light-years distant, cannot take place in less than 27 years. A round-tripper, no matter how quickly the journey has seemed to pass for him, will find his friends on earth a minimum of 54 years older. In the same way, a trip to the galaxy of Andromeda cannot take less than 2,100,000 years of objective time, Andromeda being 2,100,000 light-years distant. Once again, time and distance are simultaneously expressed.

My idea, then, is to apply this same principle to the realm of ultra-short intervals of time.

Instead of concentrating on the tremendously long distances light can cover in ordinary units of time, why not concentrate on the tremendously short times required for light to cover ordinary units of distance?

If we're going to speak of a light-second as equal to the distance covered by light (in a vacuum) in one second and set it equal to 186,273 miles, why not speak of a "light-mile" as equal to the time required for light (in a vacuum) to cover a distance of one mile, and set that equal to 1/186,273 seconds?

Why not, indeed? The only drawback is that 186,273 is such an uneven number. However, by a curious coincidence undreamed of by the inventors of the metric system, the speed of light is very close to 300,000 kilometers per second, so that a "light-kilometer" is equal to 1/300,000 of a second. It comes out even rounder if you note that 3⅓ light-kilometers are equal to just about 0.00001 or 10^{-5} seconds.

Furthermore, to get to still smaller units of time, it is only necessary to consider light as covering smaller and smaller distances.

Thus, one kilometer (10^5 centimeters) is equal to a million millimeters, and one millimeter (10^{-1} centimeters) is equal to a million nanometers. To go one step further down, we can say that one nanometer (10^{-7} centimeters) is equal to a million fermis. (The name "fermi" has been suggested, but has not yet been officially adopted,† as far as I know, for a unit of length equal to a millionth of a nanometer, or to 10^{-13} centimeters. It is derived, of course, from the late Enrico Fermi, and I will accept the name for the purposes of this chapter.)

So we can set up a little table of light-units for ultra-short intervals of time, beginning with a light-kilometer,†† which is itself equal to only 1/300,000 of a second.

1 light-kilometer = 1,000,000 light-millimeters
1 light-millimeter = 1,000,000 light-nanometers
1 light-nanometer = 1,000,000 light-fermis

† Since this article appeared in August 1959, the term for 10^{-13} centimeters, or 10^{-15} meters, has been accepted as "femtometer." Despite the similarity in spelling, "femto" has nothing to do with Fermi; it is from the Danish word for "fifteen."

†† If it will help, one mile equals 1⅗ kilometers, and one inch equals 25½ millimeters.

To relate these units to conventional units of time, we need only set up another short table:

3⅓ light-kilometers	=	10^{-5} seconds (i.e. a hundred-thousandth of a second)
3⅓ light-millimeters	=	10^{-11} seconds (i.e. a hundred-billionth of a second)
3⅓ light-nanometers	=	10^{-17} seconds (i.e. a hundred-quadrillionth of a second)
3⅓ light-fermis	=	10^{-23} seconds (i.e. a hundred-sextillionth of a second)

But why stop at the light-fermi? We can proceed on downward, dividing by a million indefinitely.

Consider the fermi again. It is equal to 10^{-13} centimeters, a ten-trillionth of a centimeter. What is interesting about this particular figure, and why the name of an atomic physicist should have been suggested for the unit, is that 10^{-13} centimeters is also the approximate diameter of the various subatomic particles.

A light-fermi, therefore, is the time required for a ray of light to travel from one end of a proton to the other. The light-fermi is the time required for the fastest known motion to cover the smallest tangible distance. Until the day comes that we discover something faster than the speed of light or something smaller than subatomic particles, we are not likely ever to have to deal with an interval of time smaller than the light-fermi. As of now, the light-fermi is the ultimate split of the second.

Of course, you may wonder what can happen in the space of a light-fermi. And if something did happen in that unimaginably small interval, how could we tell it didn't take place in a light-nanometer, which is also unimaginably small for all it is equal to a million light-fermis?

Well, consider high-energy particles. These (if the energy is high enough) travel with almost the speed of light. And when one of these particles approaches another at such a speed, a reaction often takes place between them, as a result of mutual "nuclear forces" coming into play.

Nuclear forces, however, are very short-range. Their strength falls off with distance so rapidly that the forces are only appreciable within one or two fermis distance of any given particle.

We have here, then, the case of two particles passing at

the speed of light and able to interact only while within a couple of fermis of each other. It would only take them a couple of light-fermis to enter and leave that tiny zone of interaction at the tremendous speed at which they are moving. Yet reactions *do* take place!

Nuclear reactions taking place in light-fermis of time are classed as "strong interactions." They are the results of forces that can make themselves felt in the most evanescent imaginable interval, and these are the strongest forces we know of. Nuclear forces of this sort are, in fact, about 135 times as strong as the electromagnetic forces with which we are familiar.

Scientists adjusted themselves to this fact and were prepared to have any nuclear reactions involving individual subatomic particles take only light-fermis of time to transpire.

But then complications arose. When particles were slammed together with sufficient energy to undergo strong interactions, new particles not previously observed were created in the process and were detected. Some of these new particles (first observed in 1950) amazed scientists by proving to be very massive. They were distinctly more massive, in fact, than neutrons or protons, which, until then, had been the most massive particles known.

These super-massive particles are called "hyperons" (the prefix "hyper-" comes from the greek and means "over," "above," "beyond"). There are three classes of these hyperons, distinguished by being given the names of different Greek letters. There are the lambda particles, which are about 12 per cent heavier than the proton; the sigma particles, which are about 13 per cent heavier; and the xi particles, which are about 14 per cent heavier.

There were theoretical reasons for suspecting that there were one pair of lambda particles, three pairs of sigma particles, and two pairs of xi particles. These differ among themselves in the nature of their electric charge and in the fact that one of each pair is an "antiparticle." One by one, each of the hyperons was detected in bubble chamber experiments; the xi-zero particle, detected early in 1959, was the last of them. The roster of hyperons was complete.

The hyperons as a whole, however, were odd little creatures. They didn't last long, only for unimaginably small fractions of a second. To scientists, however, they seemed to last very long indeed, for nuclear forces were involved

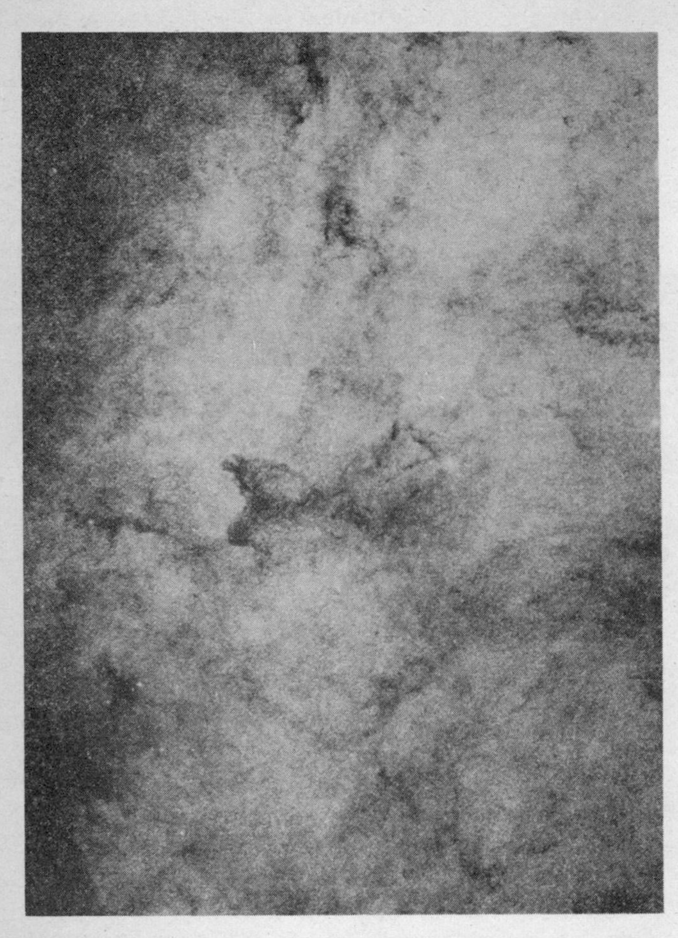

The Granger Collection

STAR CLOUDS

In ancient times, thousands and, possibly, tens of thousands were the largest number units generally needed. The arrival of larger number units, such as millions, came in connection with finance and marked the economic expansion following the discovery of the Americas and the opening of European trade with the far East.

Still larger numbers arrived through science and chiefly through astronomy. The distances of the heavenly bodies dwarfed the units of measurement that had been invented to deal with earthly objects. Even the ancients knew that the Moon was about 250,000 miles away, but by the close of the seventeenth century, it was discovered that the Sun was nearly 100,000,000 miles away and Saturn eight times that distance—and beyond them lay the stars. Really large numbers became common first of all in connection with "astronomic distances."

In the 1830s the distance of the nearer stars was determined and those turned out to be in the tens and hundreds of trillions of miles. The nearest star, Alpha Centauri, is 25,000,000,000,000 miles away and that is only the beginning. Light, which travels at a speed of 186,282 miles per second fleets across enormous distances in tiny intervals of time by Earth standards, but barely makes a dent among the stars. It takes 4.3 years for light to travel from Earth to Alpha Centauri, and that star is therefore 4.3 light-years away.

The center of our Galaxy (located beyond the vast starclouds in Sagittarius pictured here) is some 30,000 light-years away, while from end to end the Galaxy is 100,000 light-years across. The nearest giant galaxy beyond our own is the Andromeda which is 2,300,000 light-years away; the nearest quasar is over 1,000,000,000 light-years away; and the limits of the Observable Universe is some 12,000,000,000 light-years away or 70,000,000,000,000,000,000,000 miles away.

in their breakdown, which should therefore have taken place in light-fermis of time.

But they didn't. Even the most unstable of all the hyperons, the sigma-zero particle, must last at least as long as a quintillionth of a second. Put that way, it sounds like a satisfactorily short period of time—not long enough to get really bored. But when the interval is converted from conventional units to light-units, we find that a quintillionth of a second is equal to 30,000 light-fermis.

Too long!

And even so, 30,000 light-fermis represent an extraordinarily short lifetime for a hyperon. The others, including the recently discovered xi-zero particle, have half-lives of about 30,000,000,000,000 light-fermis, or 30 light-millimeters.

Since the nuclear forces bringing about the breakdown of hyperons last at least ten trillion times as long an interval of time as that required to form them, those forces must be that much weaker than those involved in the "strong interactions." Naturally, the new forces are spoken of as being involved in the "weak interactions"; and they are weak indeed, being almost a trillion times weaker than even electromagnetic forces.

In fact, the new particles which were involved in "weak interactions" were called "strange particles," partly because of this, and the name has stuck. Every particle is now given a "strangeness number" which may be +1, 0, −1 or −2. Ordinary particles such as protons and neutrons have strangeness numbers of 0; lambda and sigma particles have strangeness numbers of −1, xi particles have strangeness numbers of −2, and so forth. Exactly what the strangeness number signifies is not yet completely clear; but it can be worked with now and figured out later.

The path and the activities of the various hyperons (and of the other subatomic particles as well) are followed by their effects upon the molecules with which they collide. Such a collision usually involves merely the tearing off of an electron or two from the air molecules. What is left of the molecule is a charged "ion."

An ion is much more efficient as a center about which a water droplet can form, than is the original uncharged molecule. If a speeding particle collides with molecules in a sample of air which is supercharged with water vapor (as in a Wilson cloud chamber) or in liquid hydrogen at the point of boiling (as in a bubble chamber), each ion that is produced is immediately made the center of a water droplet or gas droplet, respectively. The moving particle marks its path, therefore, with a delicate line of water drops. When the particle breaks down into two other particles, moving off in two different directions, the line of water gives away by splitting into a Y form.

It all happens instantaneously to our merely human senses. But photograph upon photograph of the tracks that result will allow nuclear physicists to deduce the chain of events that produced the different track patterns.

Only subatomic particles that are themselves charged are very efficient in knocking electrons out of the edges of air molecules. For that reason, only charged particles can be followed by the water traceries. And, also for that reason, in any class of particles, the uncharged or neutral varieties are the last to be detected.

For instance, the neutron, which is uncharged, was not discovered until eighteen years after the discovery of the similar, but electrically charged proton. And in the case of the hyperons, the last to be found was xi-zero, one of the uncharged varieties. (The "zero" means "zero charge.")

Yet the uncharged particles can be detected by the very absence of a trace. For instance, the xi-zero particle was formed from a charged particle, and broke down, eventually, into another type of charged particle. In the photograph that finally landed the jackpot (about seventy thousand were examined), there were lines of droplets separated by a significant gap! That gap could not be filled by any known uncharged particle, for any of those would have brought about a different type of gap or a different sequence of events at the conclusion of the gap. Only the xi-zero could be made to fit; and so, in this thoroughly negative manner, the final particle was discovered.

And where do the light-units I'm suggesting come in? Why, consider that a particle traveling at almost the speed of light has a chance, if its lifetime is about 30 light-millimeters, to travel 30 light-millimeters before breaking down.

The one implies the other. By using conventional units, you might say that a length of water droplets of about 30 millimeters implies a half-life of about a trillionth of a second (or vice versa), but there is no obvious connection between the two numerical values. To say that a track of 30 millimeters implies a half-life of 30 light-millimeters is equally true, and how neatly it ties in. Once again, as in the case of astronomical distances, the use of the speed of light allows one number to express both distance and time.

A group of particles which entered the scene earlier than the hyperons are the "mesons." These are middleweight particles, lighter than protons or neutrons, but heavier than electrons. (Hence their name, from a Greek word meaning "middle.")

There are three known varieties of these particles, too. The two lighter varieties are also distinguished by means of Greek letters. They are the mu-mesons, discovered in 1935, which are about 0.11 as massive as a proton, and the pi-mesons, discovered in 1947, which are about 0.15 as massive as protons. Finally, beginning in 1949, various species of unusually heavy mesons, the K-mesons, were discovered. These are about 0.53 as massive as protons.

On the whole, the mesons are less unstable than the hyperons. They have longer half-lives. Whereas even the most stable of the hyperons has a half-life of only 30 light-millimeters, the meson half-lives generally range from that value up through 8,000 light-millimeters for those pi-mesons carrying an electric charge, to 800,000 light-millimeters for the mu-mesons.

By now, the figure of 800,000 light-millimeters ought to give you the impression of a long half-life indeed, so I'll just remind you that by conventional units it is the equivalent of 1/400,000 of a second.

A short time to us, but a long, lo-o-o-ong time on the nuclear scale.

Of the mesons, it is only the K-variety that comes under the heading of strange particles. The K-plus and K-zero mesons have a strangeness number of +1, and the K-minus meson, a strangeness number of −1.

The weak interactions, by the way, recently opened the door to a revolution in physics. For the first eight years or so after their discovery, the weak interactions had seemed to be little more than confusing nuisances. Then, in 1957, as a result of studies involving them, the "law of conservation of parity" was shown not to apply to all processes in nature.

I won't go into the details of that, but it's perhaps enough to say that the demonstration thunderstruck physicists; that the two young Chinese students who turned the trick (the older one was in his middle thirties) were promptly awarded the Nobel Prize; and that a whole new horizon in nuclear theory seems to be opening up as a result.

Aside from the mesons and hyperons, there is only one unstable particle known—the neutron.* Within the atomic nucleus the neutron is stable; but in isolation, it eventually

* Soon after this chapter first appeared, physicists began to detect numbers of "resonance particles" with half-lives in the order of 10^{-23} seconds. For them a light-fermi is a life's journey.

breaks down to form a proton, an electron and a neutrino. (Of course, antiparticles such as positrons and antiprotons are unstable in the sense that they will react with electrons and protons, respectively. Under ordinary circumstances this will happen in a millionth of a second or so. However, if these antiparticles were in isolation, they would remain as they were indefinitely, and that is what we mean by stability.)

The half-life of the neutron breakdown is 1010 seconds (or about 17 minutes), and this is about a billion times longer than the half-life of the breakdown of any other known particle.

In light-units, the half-life of the neutron would be 350,000,000 light-kilometers. In other words, if a number of neutrons were speeding at the velocity of light, they would travel 350,000,000 kilometers (from one extreme of earth's orbit to the other, plus a little extra) before half had broken down.

Of course, neutrons as made use of by scientists don't go at anything like the speed of light. In fact, the neutrons that are particularly useful in initiating uranium fission are very slow-moving neutrons that don't move any faster than air molecules do. Their speed is roughly a mile a second.

Even at that creep, a stream of neutrons would travel a thousand miles before half had broken down. And in that thousand miles, many other things have a chance to happen to them. For instance, if they're traveling through uranium or plutonium, they have a chance to be absorbed by nuclei and to initiate fission. And to help make the confusing and dangerous—but exciting—world we live in today.

Eleven
THE HEIGHT OF UP

Most of us would consider the surface of the sun to be pretty hot. Its temperature, as judged by the type of radiation it emits, is about 6000° K. (with "K." standing for the Kelvin scale of temperature).

However, Homo sapiens, with his own hot little hands, can do better than that. He has put together nuclear fission bombs which can easily reach temperatures well beyond 100,000° K.

To be sure, though, nature isn't through. The sun's corona has an estimated temperature of about 1,000,000° K., and the center of the sun is estimated to have a temperature of about 20,000,000° K.

Ah, but man can top that, too. The hydrogen bomb develops temperatures of about 100,000,000° K.

And yet nature still beats us, since it is estimated that the central regions of the very hottest stars (the sun itself is only a middling warm one) may reach as high as 2,000,000,000° K.

Now two billion degrees is a tidy amount of heat even when compared to a muggy day in New York, but the questions arise: How long can this go on? Is there any limit to how hot a thing can be?

Or to put it another way, How hot is hot?

That sounds like asking, How high is up? and I wouldn't do such a thing except that our twentieth century has seen the height of upness scrupulously defined in some respects.

For instance, in the good old days of Newtonian physics there was no recognized limit to velocity. The question, How fast is fast? had no answer. Then along came Einstein, who advanced the notion, now generally accepted, that the maximum possible velocity is that of light, which is equal to 186,274 miles per second, or, in the metric system, 299,776 kilometers per second. *That* is the fastness of fast.

So why not consider the hotness of hot?

One of the reasons I would like to do just that is to take

up the question of the various temperature scales and their interconversions for the general edification of the readers. The subject now under discussion affords an excellent opportunity for just that.

For instance, why did I specify the Kelvin scale of temperature in giving the figures above? Would there have been a difference if I had used Fahrenheit? How much and why? Well, let's see.

The measurement of temperature is a modern notion, not more than 350 years old. In order to measure temperature, one must first grasp the idea that there are easily observed physical characteristics which vary more or less uniformly with change in the subjective feeling of "hotness" and "coldness." Once such a characteristic is observed and reduced to quantitative measurement, we can exchange a subjective, "Boy, it's getting hotter," to an objective, "The thermometer has gone up another three degrees."

One applicable physical characteristic, which must have been casually observed by countless people, is the fact that substances expand when warmed and contract when cooled. The first of all those countless people, however, who tried to make use of this fact to measure temperature was the Italian physicist Galileo Galilei. In 1603 he inverted a tube of heated air into a bowl of water. As the air cooled to room temperature, it contracted and drew the water up into the tube. Now Galileo was ready. The water level kept on changing as room temperature changed, being pushed down when it warmed and expanded the trapped air, and being pulled up when it cooled and contracted the trapped air. Galileo had a thermometer (which, in Greek, means "heat measure"). The only trouble was that the basin of water was open to the air and air pressure kept changing. That also showed the water level up and down, independently of temperature, and queered the results.

By 1654, the Grand Duke of Tuscany, Ferdinand II, evolved a thermometer that was independent of air pressure. It contained a liquid sealed into a tube, and the contraction and expansion of the liquid itself was used as an indication of temperature change. The volume change in liquids is much smaller than in gases, but by using a sizable reservoir of liquid which was filled so that further expansion could only take place up a very narrow tube, the rise and fall within that tube, for even tiny volume changes, was considerable.

This was the first reasonably accurate thermometer, and was also one of the few occasions on which the nobility contributed to scientific advance.

With the development of a desire for precision, there slowly arose the notion that, instead of just watching the liquid rise and fall in the tube, one ought to mark off the tube at periodic intervals so that an actual quantitative measure could be made. In 1701, Isaac Newton suggested that the thermometer be thrust into melting ice and that the liquid level so obtained be marked as 0, while the level attained at body temperature be marked off as 12, and the interval divided into twelve equal parts.

The use of a twelve-degree scale for this temperature range was logical. The English had a special fondness for the duodecimal system (and need I say that Newton was English?). There are twelve inches to the foot, twelve ounces to the Troy pound, twelve shillings to the pound, twelve units to a dozen and twelve dozen to a gross. Why not twelve degrees to a temperature range? To try to divide the range into a multiple of twelve degrees—say into twenty-four or thirty-six degrees—would carry the accuracy beyond that which the instrument was then capable of.

But then, in 1714, a German physicist named Gabriel Daniel Fahrenheit made a major step forward. The liquid that had been used in the early thermometers was either water or alcohol. Water, however, froze and became useless at temperatures that were not very cold, while alcohol boiled and became useless at temperatures that were not very hot. What Fahrenheit did was to substitute mercury. Mercury stayed liquid well below the freezing point of water and well above the boiling point of alcohol. Furthermore, mercury expanded and contracted more uniformly with temperature than did either water or alcohol. Using mercury, Fahrenheit constructed the best thermometers the world had yet seen.

With his mercury thermometer, Fahrenheit was now ready to use Newton's suggestion; but in doing so, he made a number of modifications. He didn't use the freezing point of water for his zero (perhaps because winter temperatures below that point were common enough in Germany and

Fahrenheit wanted to avoid the complication of negative temperatures). Instead, he set zero at the very lowest temperature he could get in his laboratory, and that he attained by mixing salt and melting ice.

Then he set human body temperature at 12, following Newton, but that didn't last either. Fahrenheit's thermometer was so good that a division into twelve degrees was unnecessarily coarse. Fahrenheit could do eight times as well, so he set body temperature at 96.

On this scale, the freezing point of water stood at a little under 32, and the boiling point at a little under 212. It must have struck him as fortunate that the difference between the two should be about 180 degrees, since 180 was a number that could be divided evenly by a large variety of integers including 2, 3, 4, 5, 6, 9, 10, 12, 15, 18, 20, 30, 36, 45, 60 and 90. Therefore, keeping the zero point as was, Fahrenheit set the freezing point of water at exactly 32 and the boiling point at exactly 212. That made body temperature come out (on the average) at 98.6°, which was an uneven value, but this was a minor point.

Thus was born the Fahrenheit scale, which we, in the United States, use for ordinary purposes to this day. We speak of "degrees Fahrenheit" and symbolize it as "° F." so that the normal body temperature is written 98.6° F.

In 1742, however, the Swedish astronomer Anders Celsius, working with a mercury thermometer, made use of a different scale. He worked downward, setting the boiling point of water equal to zero and the freezing point at 100. The next year this was reversed because of what seems a natural tendency to let numbers increase with increasing heat and not with increasing cold.

Because of the hundredfold division of the temperature range in which water was liquid, this is called the Centigrade scale from Latin words meaning "hundred steps." It is still common to speak of measurements on this scale as "degrees Centigrade," symbolized as "° C." However, a couple of years back, it was decided, at an international conference, to call this scale after the inventor, following the Fahrenheit precedent. Officially, then, one should speak of the "Celsius scale" and of "degrees Celsius." The symbol remains "° C."

The Celsius scale won out in most of the civilized world. Scientists, particularly, found it convenient to bound the liquid range of water by 0° at the freezing end and 100° at the boiling end. Most chemical experiments are conducted in water, and a great many physical experiments, involving heat, make use of water. The liquid range of water is therefore the working range, and as scientists were getting used to forcing measurements into line with the decimal system (soon they were to adopt the metric system which is decimal throughout), 0 and 100 were just right. To divide the range between 0 and 10 would have made the divisions too coarse, and division between 0 and 1000 would have been too fine. But the boundaries of 0 and 100 were just right.

However, the English had adopted the Fahrenheit scale. They stuck with it and passed it on to the colonies which, after becoming the United States of America, stuck with it also.

Of course, part of the English loyalty was the result of their traditional traditionalism, but there was a sensible reason, too. The Fahrenheit scale is peculiarly adapted to meteorology. The extremes of 0 and 100 on the Fahrenheit scale are reasonable extremes of the air temperature in western Europe. To experience temperatures in the shade of less than 0° F. or more than 100° F. would be unusual indeed. The same temperature range is covered on the Celsius scale by the limits −18° C. and 38° C. These are not only uneven figures but include the inconvenience of negative values as well.

So now the Fahrenheit scale is used in English-speaking countries and the Celsius scale everywhere else (including those English-speaking countries that are usually not considered "Anglo-Saxon"). What's more, scientists everywhere, *even* in England and the United States, use the Celsius scale.

If an American is going to get his weather data thrown at him in degrees Fahrenheit and his scientific information in degrees Celsius, it would be nice if he could convert one into the other at will. There are tables and graphs that will do it for him, but one doesn't always carry a little table or graph on one's person. Fortunately, a little arithmetic is all that is really required.

In the first place, the temperature range of liquid water is covered by 180 equal Fahrenheit degrees and also by 100 equal Celsius degrees. From this, we can say at once that 9 Fahrenheit degrees equal 5 Celsius degrees. As a first approximation, we can then say that a number of Celsius degrees multiplied by $\frac{9}{5}$ will give the equivalent number of Fahrenheit degrees. (After all, 5 Celsius degrees multiplied by $\frac{9}{5}$ does indeed give 9 Fahrenheit degrees.)

Now how does this work out in practice? Suppose we are speaking of a temperature of 20° C., meaning by that a temperature that is 20 Celsius degrees above the freezing point of water. If we multiply 20 by $\frac{9}{5}$ we get 36, which is the number of Fahrenheit degrees covering the same range; the range, that is, above the freezing point of water. But the freezing point of water on the Fahrenheit scale is 32°. To say that a temperature is 36 Fahrenheit degrees above the freezing point of water is the same as saying it is 36 plus 32 or 68 Fahrenheit degrees above the Fahrenheit zero; and it is degrees above zero that is signified by the Fahrenheit reading. What we have proved by all this is that 20° C. is the same as 68° F. and vice versa.

This may sound appalling, but you don't have to go through the reasoning each time. All that we have done can be represented in the following equation, where F represents the Fahrenheit reading and C the Celsius reading:

$$F = \frac{9}{5}C + 32 \qquad \text{(Equation 15)}$$

To get an equation that will help you convert a Fahrenheit reading into Celsius with a minimum of thought, it is only necessary to solve Equation 1 for C, and that will give you:

$$C = \frac{5}{9}(F - 32) \qquad \text{(Equation 16)}$$

To give an example of the use of these equations, suppose, for instance, that you know that the boiling point of ethyl alcohol is 78.5° C. at atmospheric pressure and wish to know what the boiling point is on the Fahrenheit scale. You need only substitute 78.5 for C in Equation 15. A little arithmetic and you find your answer to be 173.3° F.

And if you happen to know that normal body temperature is 98.6° F. and want to know the equivalent in Celsius, it is only necessary to substitute 98.6 for F in Equation 16. A little arithmetic again, and the answer is 37.0° C.

But we are not through. In 1787, the French chemist Jacques Alexandre César Charles discovered that when a gas was heated, its volume expanded at a regular rate, and that when it was cooled, its volume contracted at the same rate. This rate was 1/273 of its volume at 0° C. for each Celsius degree change in temperature.

The expansion of the gas with heat raises no problems, but the contraction gives rise to a curious thought. Suppose a gas has the volume of 273 cubic centimeters at 0° C. and it is cooled. At −1° C. it has lost 1/273 of its original volume, which comes to 1 cubic centimeter, so that only 272 cubic centimeters are left. At −2° C. it has lost another 1/273 of its original volume and is down to 271 cubic centimeters. The perceptive reader will see that if this loss of 1 cubic centimeter per degree continues, then at −273° C., the gas will have shrunk to zero volume and will have disappeared from the face of the earth.

Undoubtedly, Charles and those after him realized this, but didn't worry. Gases on cooling do not, in actual fact, follow Charles's law (as this discovery is now called) exactly. The amount of decrease slowly falls off and before the −273° point is reached, all gases (as was guessed then and as is known now) turn to liquids, anyway; and Charles's law does not apply to liquids. Of course, a "perfect gas" may be defined as one for which Charles's law works perfectly. A perfect gas would indeed contract steadily and evenly, would never turn to liquid, and would disappear at −273°. However, since a perfect gas is only a chemist's abstraction and can have no real existence, why worry?

Slowly, through the first half of the nineteenth century, however, gases came to be looked upon as composed of discrete particles called molecules, all of which were in rapid and random motion. The various particles therefore possessed kinetic energy (i.e. "energy of motion"), and temperature came to be looked upon as a measure of the kinetic energy of the molecules of a substance under given conditions. Temperature and kinetic energy rise and fall together. Two substances are at the same temperature when the molecules of each have the same kinetic energy. In fact, it is the equality of kinetic energy which our human senses (and our nonhuman thermometers) register as "being of equal temperature."

The individual molecules in a sample of gas do not all possess the same energies, by any means, at any given temperature. There is a large range of energies which are produced by the effect of random collisions that happen to give some molecules large temporary supplies of energy, leaving others with correspondingly little. Over a period of time and distributed among all the molecules present, however, there is an "average kinetic energy" for every temperature, and this is the same for molecules of all substances.

In 1860, the Scottish mathematician Clerk Maxwell worked out equations which expressed the energy distribution of gas molecules at any temperature and gave means of calculating the average kinetic energy.

Shortly after, a British scientist named William Thomson (who had just been raised to the ranks of the nobility with the title of Baron Kelvin) suggested that the kinetic energy of molecules be used to establish a temperature scale. At 0° C. the average kinetic energy per molecule of any substance is some particular value. For each Celsius degree that the temperature is lowered, the molecules lose 1/273 of their kinetic energy. (This is like Charles's law, but whereas the decrease of gas volume is not perfectly regular, the decrease in molecular energies—of which the decrease in volume is only an unavoidable and imperfect consequence —*is* perfectly regular.) This means that at −273° C., or, more exactly, at −273.16° C., the molecules have zero kinetic energy. The substance—any substance—can be cooled no further, since negative kinetic energy is inconceivable.

The temperature of −273.16° C. can therefore be considered an "absolute zero." It a new scale is now invented in which absolute zero is set equal to 0° and the size of the degree is set equal to that of the ordinary Celsius degree, then any Celsius reading could be converted to a corresponding reading on the new scale by the addition of 273.16 (The new scale is referred to as the absolute scale or, more appropriately in view of the convention that names scales after the inventors, the Kelvin scale, and degrees on this scale can be symbolized as either "° A." or "° K.") Thus, the freezing point of water is 273.16° K. and the boiling point of water is 373.16° K.

In general:

$$K = C + 273.16 \qquad \text{(Equation 17)}$$
$$C = K - 273.16 \qquad \text{(Equation 18)}$$

You might wonder why anyone would need the Kelvin scale. What difference does it make just to add 273.16 to every Celsius reading? What have we gained? Well, a great many physical and chemical properties of matter vary with temperature. To take a simple case, there is the volume of a perfect gas (which is dealt with by Charles's law). The volume of such a gas, at constant pressure, varies with temperature. It would be convenient if we could say that the variation was direct; that is, if doubling the temperature meant doubling the volume.

If, however, we use the Celsius scale, we cannot say this. If we double the temperature from, say, 20° C. to 40° C., the volume of the perfect gas does *not* double. It increases by merely one-eleventh of its original volume. If we use the Kelvin scale, on the other hand, a doubling of temperature does indeed mean a doubling of volume. Raising the temperature from 20° K. to 40° K., then to 80°K., then to 160° K., and so on, will double the volume each time.

In short, the Kelvin scale allows us to describe more conveniently the manner in which the universe behaves as temperature is varied—more conveniently than the Celsius scale, or any scale with a zero point anywhere but at absolute zero, can.

Another point I can make here is that in cooling any substance, the physicist is withdrawing kinetic energy from its molecules. Any device ever invented to do this only succeeds in withdrawing a fraction of the kinetic energy present, however little the amount present may be. Less and less energy is left as the withdrawal step is repeated over and over, but the amount left is never zero.

For this reason, scientists have not reached absolute zero and do not expect to, although they have done wonders and reached a temperature of 0.00001° K.

At any rate, here is another limit established, and the question: How cold is cold? is answered.

But the limit of cold is a kind of "depth of down" as far as temperature is concerned, and I'm after the "height of up," the question of whether there is a limit to hotness and, if so, where it might be.

Let's take another look at the kinetic energy of molecules. Elementary physics tells us that the kinetic energy (E) of a moving particle is equal to $\frac{1}{2}mv^2$, where "m" represents the mass of a particle and "v" its velocity. If we solve the equation $E = \frac{1}{2}mv^2$ for "v", we get:

$$v = \sqrt{\frac{2E}{m}} = 1.414\sqrt{\frac{E}{m}} \quad \text{(Equation 19)}$$

But the kinetic energy content is measured by the temperature (T), as I've already said. Consequently, we can substitute "T" for "E" in Equation 19 (and I will also change the numerical constant to allow the figures to come out correctly in the units I will use). We can say that:

$$v = 0.158\sqrt{\frac{T}{m}} \quad \text{(Equation 20)}$$

Now then, if in Equation 20 the temperature (T) is given in degrees Kelvin, and the mass (m) of the particle is given in atomic units, then the average velocity (v) of the particles will come out in kilometers per second. (If the numerical constant were changed from 0.158 to 0.098, the answer would come out in miles per second.)

For instance, consider a sample of helium gas. It is composed of individual helium atoms, each with a mass of 4, in atomic units. Suppose the temperature of the sample is the freezing point of water (273° K.). We can therefore substitute 273 for "T" and 4 for "m" in Equation 20. Working out the arithmetic, we find that the average velocity of helium atoms at the freezing point of water is 1.31 kilometers per second (0.81 miles per second).

This will work out for other values of "T" and "m." The velocity of oxygen molecules (with a mass of 32) at room temperature (300° K.) works out as $0.158\sqrt{300/32}$ or 0.48 kilometers per second. The velocity of carbon dioxide molecules (with a mass of 44) at the boiling point of water (373° K.) is 0.46 kilometers per second, and so on.

Equation 20 tells us that at any given temperature, the lighter the particle the faster it moves. It also tells us that at absolute zero (where T = 0) the velocity of any atom or molecule, whatever its mass, is zero. This is another way of looking at the absoluteness of absolute zero. It is the point of absolute (well, almost absolute) atomic or molecular rest.

The Granger Collection

WILLIAM THOMSON, 1ST BARON KELVIN

Lord Kelvin was born William Thomson, on June 26, 1824, in Belfast, Ireland. He was the son of an eminent mathematician and was an infant prodigy who attended his father's lectures with delight when only eight years old. At eleven he entered the University of Glasgow and his first paper on mathematics was read to the Royal Society of Edinburgh by an aged professor since it seemed undignified for the staid assemblage to be lectured to by a schoolboy.

When Kelvin began to teach, he was one of the first to use the laboratory as well as the lecture hall. His first laboratory was a converted wine cellar. Interested in the problem of heat, Kelvin was one of the founders of thermodynamics (a word he introduced) and was the first to make it clear that there was a limit to coldness. When atoms and molecules lost all their kinetic energy, they had reached the "absolute zero" and Kelvin worked out a temperature scale based on this.

During the years when engineers were trying to lay the Atlantic cable, in order to make possible transoceanic telegraphy for the first time, Thomson studied the capacity of a cable to carry an electric signal and invented improvements without which the Atlantic cable would not have worked. In 1866 he was knighted because of his achievements in this respect, and in 1892 he was made a baron.

He was the first to introduce the telephone into Great Britain and made a number of important inventions in connection with oceanography. He died near Largs, Scotland, on December 17, 1907, in his eighty-fourth year, having lived long enough to witness the great changes that came with the discovery of radioactivity, but, in his old age, unwilling to accept them.

But if a velocity of zero is a lower limit, is there not an upper limit to velocity as well? Isn't this upper limit the velocity of light, as I mentioned at the beginning of the article? When the temperature goes so high that "v" in Equation 6 reaches the speed of light and can go no higher, have we not reached the absolute height of up, the ultimate hotness of hot? Let's suppose all that is so, and see where it leads us.

Let's begin by solving Equation 6 for "T." It comes out:

$$T = 40\ mv^2 \qquad \text{(Equation 21)}$$

The factor, 40, only holds when we use units of degrees Kelvin, and kilometers per second.

Let's set the value of "v" (the molecular velocity) equal to the maximum possible, or the 299,776 kilometers per second which is the velocity of light. When we do that, we get what would seem to be the maximum possible temperature (T_{max}):

$$T_{max} = 3,600,000,000,000 \text{ m} \quad \text{(Equation 22)}$$

But now we must know the value of "m" (the mass of the particles involved). The higher the value of "m," the higher the maximum temperature.

Well, at temperatures in the millions all molecules and atoms have broken down to bare nuclei. At temperatures of hundreds of millions and into the low billions, fusion reactions between simple nuclei are possible so that complicated nuclei can be built up. At still higher temperatures, this must be reversed and all nuclei must break apart into simple protons and neutrons.

Let's suppose, then, that in the neighborhood of our maximum possible temperature, which is certainly over a trillion degrees, only protons and neutrons can exist. These have a mass of 1 on the atomic scale. Consequently, from Equation 22, we must conclude that the maximum possible temperature is 3,600,000,000,000° K.

Or must we?

For alas, I must confess that in all my reasoning from Equation 19 on there has been a fallacy. I have assumed that the value of "m" is constant; that if a helium atom has a mass of 4, it has a mass of 4 under all conceivable circumstances. This would be so, as a matter of fact, if the Newtonian view of the universe were correct, but in the Newtonian universe there is no such thing as a maximum velocity and therefore no upper limit to temperature.

On the other hand, the Einsteinian view of the universe, which gives an upper limit of velocity and therefore seems to offer the hope of an upper limit of heat, does not consider mass a constant. The mass of any object (however small under ordinary conditions, as long as it is greater than zero) increases as its velocity increases, becoming indefinitely large as one gets closer and closer to the velocity of light. (A shorthand way of putting this is: "Mass becomes infinite at the velocity of light.") At ordinary velocities, say of no more than a few thousand kilometers per second, the increase in mass is quite small and need not be worried about except in the most refined calculations.

However, when we are working near the velocity of light

or even at it as I was trying to do in Equation 22, "m" becomes very large and reaches toward the infinite regardless of the particle being considered, and so consequently does "T_{max}." There is no maximum possible temperature in the Einsteinian universe any more than in the Newtonian. In this particular case, there is no definite height to up.*

* This chapter first appeared in October 1959. It is purely ivory-tower speculation, but it got a reader to thinking. The reader was Hong Yee Chiu, a theoretical physicist who grew curious as to what the highest *real* temperature might be. This led him to calculations of nuclear reactions in the center of stars, to suggestions as to the cause of star-implosions and super-novae, to the role played by neutrino formation in the process, and, in general, to the founding of "neutrino astronomy." (If I can't do great things myself, I like to contribute to the doing of them by others.)

Twelve
ORDER! ORDER!

One of the big dramatic words in science is ENTROPY. It comes so trippingly and casually to the tongue; yet when the speaker is asked to explain the term, lockjaw generally sets in. Nor do I exonerate myself in this respect. I, too, have used the word with fine abandon and have learned to change the subject deftly when asked to explain its meaning.

But I must not allow myself to be cowardly forever. So, with lips set firmly and with face a little pale, here I go. . . .

I will begin with the law of conservation of energy. This states that energy may be converted from one form to another but can be neither created nor destroyed.

This law is an expression of the common experience of mankind. That is, no one knows any reason *why* energy cannot be created or destroyed; it is just that so far neither the most ingenious experiment nor the most careful observation has ever unearthed a case where energy is either created or destroyed.

The law of conservation of energy was established in the 1840s, and it rocked along happily for half a century. It was completely adequate to meet all earthly problems that came up. To be sure, astronomers wondered whence came the giant floods of energy released by the sun throughout the long history of the solar system, and could find no answer which satisfactorily met the requirements of both astronomy and the law of conservation of energy.

However, that was the sun. There were no problems on earth—until radioactivity was discovered.

In the 1890s the problem arose of finding out from whence came the tremendous energy relased by radioactive substances. For a decade or so, the law of conservation of energy looked sick indeed. Then, in 1905, Albert Einstein demonstrated (mathematically) that mass and energy had to be different forms of the same thing, and that a very tiny bit of mass was equivalent to a great deal of energy. All the energy released by radioactivity was at the expense of the disappearance of an amount of mass too small to measure by ordinary methods. And in setting up this proposition, a source of energy was found which beautifully explained the radiation of the sun and other stars.

In the years after 1905, Einstein's theory was demonstrated experimentally over and over again, with the first atom bomb of 1945 as the grand culmination. The law of conservation of energy is now more solidly enthroned than ever, and scientists do not seriously expect to see it upset at any time in the future except under any but the most esoteric circumstances conceivable.

In fact, so solidly is the law enthroned that no patent office in its right mind would waste a split-second considering any device that purported to deliver more energy than it consumed. (This is called a "perpetual motion machine of the first class.")

The first machine that converted heat into mechanical work on a large scale was the steam engine, invented at the beginning of the eighteenth century by Thomas Newcomen and made thoroughly practical at the end of that century by James Watt. Since the steam engine produced work through the movement of energy in the form of heat from a hot reservoir of steam to a cold reservoir of water, the study of the interconversion of energy and work was named "thermodynamics" from Greek words meaning "movement of heat." The law of conservation of energy is so fundamental to devices such as the steam engine, that the law is often called the First Law of Thermodynamics.

The First Law tells us that if the reservoir of steam contains a certain amount of energy, you cannot get more work out of a steam engine than is equivalent to that much energy. That seems fair enough, perhaps; you can't get something for nothing.

But surely you can at least get out the full amount of work equivalent to the energy content of the steam, at least assuming you cut down on waste and on friction.

Alas, you can't. Though you built the most perfect steam engine possible, one without friction and without waste, you still could not convert all the energy to work. In thermodynamics, you not only can't win, you can't even break even.

The first to point this out unequivocally was a French

physicist named Sadi Carnot, in 1824. He stated that the fraction of the heat energy that could be converted to work even under ideal conditions depended upon the temperature difference between the hot reservoir and the cold reservoir. The equation he presented is this:

$$\text{ideal efficiency} = \frac{T_2 - T_1}{T_2}$$

where T_2 is the temperature of the hot reservoir and T_1 the temperature of the cold. For the equation to make sense the temperatures should be given on the absolute scale. (The absolute scale of temperature is discussed in the previous chapter.)

If the hot reservoir is at a temperature of 400° absolute (127° C.) and the cold reservoir is at a temperature of 300° absolute (27° C.), then ideal efficiency is:

$$\frac{400-300}{400}$$

or exactly 0.25. In other words, one-quarter of the heat content of the steam, *at best*, can be converted into work, while the other three-quarters is simply unused.

Furthermore, if you had only a hot reservoir and nothing else, so that it had to serve as hot and cold reservoir both, Carnot's equation would give the ideal efficiency as:

$$\frac{400-400}{400}$$

which is exactly zero. The steam has plenty of energy in it, but none of that energy, none at all, can be converted to work unless somewhere in this device there is a temperature difference.

An analogous situation exists in other forms of energy, and the situation may be easier to understand in cases more mundane than the heat engine. A large rock at the edge of a cliff can do work, provided it moves from its position of high gravitational potential to one of low gravitational potential, say at the bottom of the cliff. The smaller the difference in gravitational potentials (the lower the cliff), the less work the rock can be made to do in falling. If there is no cliff at all, but simply a plateau of indefinite extent, the rock cannot fall, and can do no work, even though the plateau may be six miles high.

We can say then: No device can deliver work out of a system at a single energy potential level.

This is one way of stating what is called the Second Law of Thermodynamics.

A device that purports to get work out of a single energy potential level is a "perpetual motion machine of the second class." Actually almost all perpetual motion machines perpetrated by the army of misguided gadgeteers that inhabit the earth are of this type, and patent offices will waste no time on these either.

Given energy at two different potential levels, it is the common experience of mankind to observe that the energy will flow from one potential (which we can call the higher) to another (which we can call the lower) and never vice versa (unless it is pushed). In other words, heat will pass spontaneously from a hot body to a cold one; a boulder will fall spontaneously from a cliff top to a cliff bottom; an electric current will flow spontaneously from cathode to anode.

To say: "Energy will always flow from a high potential level to a low potential level" is another way of stating the Second Law of Thermodynamics. (It can be shown that an energy flow from high to low implies the fact that you cannot get work out of a one-potential system, and vice versa, so both are equivalent ways of stating the Second Law.)

Now, work is never done instantaneously. It invariably occupies time. What happens during that time? Let's suppose, for simplicity's sake, that a steam engine is functioning as a "closed system"; that is, as a sort of walled-off portion of the universe into which no energy can enter and from which no energy can depart. In such a closed-system steam engine the Second Law states that heat must be flowing from the point of high energy potential (the hot reservoir in this case) to the point of low (the cold reservoir).

As this happens, the hot reservoir cools and the cold reservoir warms. The temperature difference between hot and cold therefore decreases during the time interval over which work is being extracted. But this means that the amount of energy which can be converted to work (an amount which depends on the size of the temperature difference) must also decrease.

Conversely, the amount of energy which *cannot* be converted into work must increase. This increase in the amount of unavailable energy is the inevitable consequence of the heat flow predicted by the Second Law. Therefore to say that in any spontaneous process (one in which energy flows from high to low) the amount of unavailable energy increases with time, is just another way of stating the Second Law.

A German physicist, Rudolf Clausius, pointed all this out in 1865. He invented a quantity consisting of the ratio of heat change to temperature and called that quantity "entropy." Why he named it that is uncertain. It comes from Greek words meaning "a change in" but that seems insufficient.

In every process involving energy change, Clausius' "entropy" increases. Even if the energy levels don't approach each other with time (the cliff top and cliff bottom don't approach each other appreciably while a rock is falling), there is always some sort of resistance to the change from one energy potential to another. The falling body must overcome the internal friction of the air it falls through, the flowing electric current must overcome the resistance of the wire it flows through. In every case, the amount of energy available for conversion to work decreases and the amount of energy unavailable for work increases. And in every case this is reflected in an increase in the heat change to temperature ratio.

To be sure, we can imagine ideal cases where this doesn't happen. A hot and cold reservoir may be separated by a perfect insulator, a rock may fall through a perfect vacuum, an electric current may flow through a perfect conductor, all surfaces may be perfectly frictionless, perfectly non-radiating. In all such cases there is no entropy increase; the entropy change is zero. However such cases generally exist only in imagination; in real life the zero change in entropy may be approached but not realized. And, of course, even in the ideal case, entropy change is never negative. There is never a *decrease* in entropy.

With all this in mind, the briefest way I know to state the First and Second Laws of Thermodynamics is this:

In any closed system the total energy content remains constant while the total entropy continually increases with time.

The main line of development of the First and Second Laws took place through a consideration of heat flow alone, without regard for the structure of matter. However, the atomic theory had been announced by John Dalton in 1803, and by the mid-nineteenth century it was well-enough established for a subsidiary line of development to arise in which energy changes were interpreted by way of the movement of atoms and molecules. This introduced a statistical interpretation of the Second Law which threw a clearer light on the question of entropy decrease.

Clausius himself had worked out some of the consequences of supposing gases to be made up of randomly moving molecules, but the mathematics of such a system was brought to a high pitch of excellence by the Scottish mathematician Clerk Maxwell and the Austrian physicist Ludwig Boltzmann in 1859 and the years immediately following.

As a result of the Maxwell-Boltzmann mathematics, gases (and matter generally) could be viewed as being made up of molecules possessing a range of energies. This energy is expressed in gases as random motion, with consequent molecular collisions and rebounds, the rebounds being assumed as involving perfect elasticity so that no energy is lost in the collisions.

A particular molecule in a particular volume of gas might have, at some particular time, any amount of energy of motion ("kinetic energy") from very small amounts to very high amounts. However, over the entire volume of gas there would be some average kinetic energy of the constituent molecules, and it is this average kinetic energy which we measure as temperature.

The average kinetic energy of the molecules of a gas at 500° absolute is twice that of the molecules of a gas at 250° absolute. The kinetic energy of a particular molecule in the hot gas may be lower at a given time than that of a particular molecule in the cold gas, but the average is in direct proportion to the temperature at all times. (This is analogous to the situation in which, although the standard of living in the United States is higher than that of Egypt, a particular American may be poorer than a particular Egyptian.)

The Granger Collection

RUDOLF JULIUS EMANUEL CLAUSIUS

Clausius was born on January 2, 1822, in the town of Köslin in eastern Germany, but you won't find it on the map anymore. It is now Koszalin, Poland.

He was primarily a theoretical physicist. He contributed to the working out of the kinetic theory of gases, something which was brought to completion by Maxwell and by Ludwig Boltzmann.

In 1857, he also made some suggestions concerning the manner in which an electric current, passing through solutions, might pull molecules apart into electrically charged fragments. This was a particularly perceptive point, but it was far ahead of its time. It was not for thirty more years that this concept entered science and was accepted by it.

His most fruitful work began in 1850, however, in connection with thermodynamics. Having invented the concept of entropy and given it its name, he placed the second law of thermodynamics on a firm foundation.

His enthusiasm and brilliance was, in a way, that of a convert, for when the conservation of energy was announced as a fundamental law of physics in the late 1840s (it is the *first* law of thermodynamics) Clausius was among those who were skeptical and who hesitated to accept the notion.

By 1869, Clausius was professor of physics at the University of Bonn, a position he held for the rest of his life. In 1870 he organized a volunteer ambulance corps of Bonn students for service in the Franco-Prussian war.

He died in Bonn, on August 24, 1888.

Now suppose that a sample of hot gas is brought into contact with a sample of cold gas. The average kinetic energy of the molecules of the hot gas being higher than that of the molecules of the cold gas, we can safely suppose that in the typical case, a "hot" molecule will be moving more quickly than a "cold" molecule. (I must put "hot" and "cold" in quotation marks, because such concepts as heat and temperature do not apply to individual molecules but only to systems containing large numbers of molecules.)

When the two rebound, the total energy of the two does not change, but it may be distributed in a new way. Some varieties of redistribution will involve the "hot" molecule growing "hotter" while the "cold" molecule grows "colder" so that, in effect, the high-energy molecule gains further energy at the expense of the low-energy particle. There are many, many more ways of redistribution, however, in which the low-energy particle gains energy at the expense of the high-energy particlc, so that both end up with intermediate energies. The "hot" molecule grows "cooler" while the "cold" molecule grows "warmer."

This means that if a very large number of collisions are considered, the vast majority of those collisions will result in a more even distribution of energy. The few cases in which the energy difference becomes more extreme will be completely swamped. On the whole, the hot gas will cool, the cold gas will warm, and eventually the two gases will reach equilibrium at some single (and intermediate) temperature.

Of course, it *is* possible, in this statistical view, that through some quirk of coincidence, all the "hot" molecules (or almost all) will just happen to gain energy from all the "cold" molecules, so that heat will flow from the cold body to the hot body. This will increase the temperature difference and, consequently, the amount of energy available for conversion into work, and thus will *decrease* the unavailable energy, which we call entropy.

Statistically speaking, then, a kettle of water *might* freeze while the fire under it grows hotter. The chances of this (if worked out from Maxwell's mathematics) are so small, however, that if the entire known universe consisted of nothing but kettles of water over fires, then the probability that any single one of them would freeze during the entire known duration of the universe would be so small that we could not reasonably hope to witness even a single occurrence of this odd phenomenon if we had been watching every kettle every moment of time.

For that matter, the molecules of air in an empty stoppered bottle are moving in all directions randomly. It *is* possible that, by sheer chance, all molecules might happen to move upward simultaneously. The total kinetic energy would then be more than sufficient to overcome gravitational attraction and the bottle would spontaneously fall upward. Again the chances for this are so small that no one expects ever to see such a phenomenon.

Still it must be said that a truer way of putting the Second Law would be: "In any closed system, entropy invariably increases with time—or, at least, *almost* invariably."

It is also possible to view entropy as having something to do with "order" and "disorder." These words are hard to define in any foolproof way; but intuitively, we picture "order" as characteristic of a collection of objects that is methodically arranged according to a logical system. Where no such logical arrangement exists, the collection of objects is in "disorder."

Another way of looking at it is to say that something which is "in order" is so arranged that one part can be distinguished from another part. The less clear the distinction, the less "orderly" it is, and the more "disorderly."

A deck of cards which is arranged in suits, and according to value within each suit, is in perfect order. Any part of the deck can be distinguished from any other part. Point to the two of hearts and I will know it is the fifteenth card in the deck.

If the deck of cards is arranged in suits, but not according to value within the suits, I am less well off. I know the two of hearts is somewhere between the fourteenth and twenty-sixth cards, but not exactly where within that range. The distinction between one part of a deck and another has become fuzzier, so the deck is now less orderly.

If the cards are shuffled until there is no system that can be devised to predict which card is where, then I can tell nothing at all about the position of the two of hearts. One part of the deck cannot be distinguished from any other part and the deck is in complete disorder.

Another example of order is an array of objects in some kind of rank and file, whether it be atoms or molecules within a crystal, or soldiers marching smartly past a reviewing stand.

Suppose you were watching marching soldiers from a reviewing stand. If they were marching with perfect precision, you would see a row of soldiers pass, then a blank space, then another row, then another blank space, and so on. You could distinguish between two kinds of volumes of space within the marching columns, soldier-full and soldier-empty, in alternation.

If the soldiers fell somewhat out of step so that the lines grew ragged, the hitherto empty volumes would start containing a bit of soldier, while the soldier-filled volumes would be less soldier-filled. There would be less distinction possible between the two types of volumes, and the situation would be less orderly. If the soldiers were completely out of step, each walking forward at his own rate, all passing volumes would tend to be equally soldier-full, and the distinction would be even less and the disorder consequently more.

The disorder would not yet be complete, however. The soldiers would be marching in one particular direction, so if you could not distinguish one part of the line from another, you could at least still distinguish your own position with respect to them. From one position you would see them march to your left, from another to your right, from still another they would be marching toward you, and so on.

But if soldiers moved randomly, at any speed and in any direction; then, no matter what your position, some would be moving toward you, some away, some to your left, some to your right, and in all directions in between. You could make no distinctions among the soldiers or among your own possible positions, and disorder would be that much more nearly complete.

Now let's go back to molecules, and consider a quantity of hot gas in contact with a quantity of cold gas. If you could see the molecules of each you could distinguish the first from the second by the fact that the molecules of the first are moving, on the average, more quickly than those of the second. Without seeing the molecules, you can achieve a similar distinction by watching the mercury thread of a thermometer.

As heat flows from the hot gas to the cold gas, the difference in average molecular motion, and hence in temperature, decreases, and the distinction between the two becomes fuzzier. Finally, when the two are at the same temperature, no distinction is possible. In other words, as heat has flowed in the direction made necessary by the Second Law, disorder has increased. Since entropy has also increased, we might wonder if entropy and disorder weren't very analogous concepts.

Apparently, they are. In any spontaneous operation, entropy increases and so does disorder. If you shuffle an ordered deck of cards you get a disordered deck, but not vice versa. (To be sure, by a fiendish stroke of luck you might begin with a disordered deck and shuffle it into a perfect arrangement, but would you care to try it and see how long that takes you? And that task involves only fifty-two objects. Imagine the same thing with several quintillion quintillion and you will not be surprised that a kettle of water over a fire never grows cooler.)

Again, if soldiers in rank and file are told to break ranks, they will quickly become a disorderly mass of humans. It is extremely unlikely, on the other hand, that a disorderly mass of humans should, by sheer luck, suddenly find themselves marching in perfect rank and file.

No, indeed. As nearly as can be told, all spontaneous processes involve an increase in disorder and an increase in entropy, the two being analogous.

It can be shown that of all forms of energy, heat is the most disorderly. Consequently, in all spontaneous processes involving types of energy other than heat, some non-heat energy is always converted to heat, this in itself involving an increase in disorder and hence in entropy.

Under no actual conditions, however, can all the heat in a system be converted to some form of non-heat energy, since that in itself would imply an increase in order and hence a decrease in entropy. Instead, if some of the heat undergoes an entropy decrease and is converted to another form of energy, the remaining heat must undergo an entropy increase that more than makes up for the first change. The net entropy change over the *entire* system is an increase.

It is, of course, easy to cite cases of entropy decrease as long as we consider only parts of a system and not all of it. For instance, we see mankind extract metals from ores and devise engines of fiendish complexity out of metal ingots. We see elevators moving upwards and automobiles driving uphill, and soldiers getting into marching formation and cards placed in order. All these and a very large number of other operations involve decreases in entropy brought about by the action of life. Consequently, the feeling arises that life can "reverse entropy."

However, there is more to consider. Human beings are eating food and supporting themselves on the energy gained from chemical changes within the body. They are burning coal and oil to power machinery. They use hydro-electric power to form aluminum. In short, all the entropy-decreasing activities of man are at the expense of the entropy-increasing activities involved in food and fuel consumption, and the entropy-increasing activities far outweigh the entropy-decreasing activities. The net change is an entropy increase.

No matter how we might bang our gavels and cry out "Order! Order!" there is no order, there is only increasing disorder.

In fact, in considering entropy changes on earth, it is unfair to consider the earth alone, since our planet is gaining energy constantly from the sun. This influx of energy from the sun powers all those processes on earth that represent local decreases in entropy: the formation of coal and oil from plant life, the circulation of atmosphere and ocean, the raising of water in the form of vapor, and so on. It is for that reason we can continue to extract energy by burning oil and coal, by making use of power from wind, river currents, waterfalls, and so forth. It is all at the expense, indirectly, of the sun.

The entropy increase represented by the sun's large-scale conversion of mass to energy simply swamps the comparatively tiny entropy decreases on earth. The net entropy change of the solar system as a whole is that of a continuing huge increase with time.

Since this must be true of all the stars, nineteenth-century physicists reasoned that entropy in the universe as a whole must be increasing rapidly and that the time must come when the finite supply of energy in a finite universe must reach a state of maximum entropy.

In such a condition, the universe would no longer contain energy capable of being converted into useful work. It would be in a state of maximum disorder. It would be a homogeneous mass containing no temperature differences. There would be no changes by which to measure time, and therefore time would not exist. There would be no way of distinguishing one point in space from another and so space would not exist. Such an entropy maximum has been referred to as the "heat-death" of the universe.

But, of course, this presupposes that the universe is finite. If it were infinite, the supply of energy would be infinite and it would take all eternity for entropy to reach a maximum. Besides that, how can we be certain that the laws of thermodynamics worked out over small volumes of space in our laboratories and observed to be true (or to seem true) in the slightly larger volumes of our astronomic neighborhood, are true where the universe as a whole is concerned?

Perhaps there are processes we know nothing about as yet, which decrease entropy as quickly as it is increased by stellar activity, so that the net entropy change in the universe as a whole is zero. This might still be so even if we allow that small portions of space, such as single galaxies, might undergo continuous entropy increases and might eventually be involved in a kind of local heat-death. The theory of continuous creation does, in fact, presuppose a constant entropy level in the universe as a whole.*

And even if the universe were finite, and even if it were to reach "heat-death," would that really be the end?

Once we have shuffled a deck of cards into complete randomness, there will come an *inevitable* time, *if we wait long enough*, when continued shuffling will restore at least a partial order.

Well, waiting "long enough" is no problem in a universe at heat-death, since time no longer exists then. We can therefore be certain that after a timeless interval, the purely random motion of particles and the purely random flow of energy in a universe at maximum entropy might, here and there, now and then, result in a partial restoration of order.

It is tempting to wonder if our present universe, large as it is and complex though it seems, might not be merely the result of a very slight random increase in order over a very small portion of an unbelievably colossal universe which is virtually entirely in heat-death.

Perhaps we are merely sliding down a gentle ripple that has been set up, accidentally and very temporarily, in a quiet pond, and it is only the limitation of our own infinitesimal range of viewpoint in space and time that makes it seem to ourselves that we are hurtling down a cosmic waterfall of increasing entropy, a waterfall of colossal size and duration.

* This chapter first appeared in February 1961. The theory of continuous creation is now in disfavor, but since then the discovery of such things as quasars and black holes assures us there are odd spots in the Universe that present conditions so extreme we cannot be sure how the laws of nature would apply or what the status of entropy change would be.

Thirteen
THE MODERN DEMONOLOGY

You would think, considering my background, that had I ever so slight a chance to drag fantasy into any serious discussion of science, I would at once do so with neon lights flashing and fireworks blasting.

And yet, in the previous chapter on entropy, I completely ignored the most famous single bit of fantasy in the history of science. Yet that was only that I might devote another entire chapter to it.

When a hot body comes into contact with a cold body, heat flows spontaneously from the hot one to the cold one and the two bodies finally come to temperature equilibrium at some intermediate level. This is one aspect of the inevitable increase of entropy in all spontaneous processes involving a closed system.

In the early nineteenth century, the popular view was to consider heat a fluid that moved from hot to cold as a stone would fall from high to low. Once a stone was at the valley bottom, it moved no more. In the same way, once the two bodies reached temperature equilibrium, there could be no further heat flow under any circumstances.

In the mid-nineteenth century, however, the Scottish mathematician James Clerk Maxwell adopted the view that temperature was the measure of the average kinetic energy of the particles of a system. The particles of a hot body moved (on the average) more rapidly than did the particles of a cold body. When such bodies were in contact, the energies were redistributed. On the whole, the most probable redistribution was for the fast particles to lose velocity (and, therefore, kinetic energy) and the slow particles to gain it. In the end, the average velocity would be the same in both bodies and would be at some intermediate level.

In the case of this particle-in-motion theory, it *was* conceivable for heat flow to continue after equilibrium had been reached.

Imagine, for instance, two containers of gas connected

by a narrow passage. The entire system is at a temperature equilibrium. That is, the average energy of the molecules in any one sizable portion of it (a portion large enough to be visible in an ordinary microscope) is the same as that in any other sizable portion.

This doesn't mean that the energies of all individual molecules are equal. There are some fast ones, some very fast ones, some very very fast ones. There are also some slow ones, some very slow ones and some very very slow ones. However, they all move about higgledy-piggledy and keep themselves well scrambled. Moreover, they are also colliding among themselves millions of times a second so that the velocities and energies of any one molecule are constantly changing. Therefore, any sizable portion of the gas has its fair share of both fast and slow molecules and ends with the same temperature as any other sizable portion.

However, what if—just as a matter of chance—a number of high-energy molecules happened to move through the connecting passageway from right to left while a number of low-energy molecules happened to move through from left to right? The left container would then grow hot and the right container cold (though the average temperature overall would remain the same). A heat-flow would be set up despite equilibrium, and entropy would decrease.

Now, there is a certain infinitesimal chance, unimaginably close to zero, that this would happen through the mere random motion of molecules. The difference between "zero" and "almost-almost-almost-zero" is negligible in practice, but tremendous from the standpoint of theory; for the chance of heat-flow at equilibrium is zero in the fluid theory and almost-almost-almost-zero in the particle-in-motion theory.

Maxwell had to find some dramatic way to emphasize this difference to the general public.

Imagine, said Maxwell, that a tiny demon sat near the passage connecting the two containers of gas. Suppose he let fast molecules pass through from right to left but not vice versa. And suppose he let slow molecules through from left to right but not vice versa. In this way, fast molecules would accumulate in the left and slow ones in the right. The left half would grow hot and the right cold. Entropy would be reversed.

The demon, however, would be helpless if heat were a

The Granger Collection

JAMES CLERK MAXWELL

Maxwell, an only son, was born in Edinburgh on November 13, 1831, of a well-known Scottish family. He was an infant prodigy in mathematics and at the age of fifteen contributed a piece of original work to the Royal Society of Edinburgh that was of such quality that the members of the body flatly refused to believe the boy was the author.

He graduated from Cambridge with a second place in mathematics and in 1857 showed from mathematical considerations that Saturn's rings could not be a continuous solid body. Such a body would not be stable in Saturn's gravitational field and the rings, therefore, had to consist of numerous small, solid particles. Numerous lines of evidence have since supported his view.

In the 1860s he worked out the kinetic theory of gases and in 1871 invented the concept of "Maxwell's demon." In 1871, also, he reluctantly allowed himself to be appointed professor of experimental physics at Cambridge. He was a poor lecturer, unfortunately, consistently going over the heads of most students, so he rarely had more than three of four to speak to.

The crowning work of Maxwell's life was carried on between 1864 and 1873, when he worked out "Maxwell's equations." These completely expressed the observed facts concerning electricity and magnetism, showed that the two were separate aspects of the electromagnetic field, showed also that light was an electromagnetic radiation and that there must be similar radiations of all wavelengths. These equations survived the revolutionary appearance of quantum theory and relativity, and remain as valid today as ever.

Maxwell died of cancer in Cambridge on November 5, 1879, just before his forty-eighth birthday.

continuous fluid—and in this way Maxwell successfully dramatized the difference in theories.

Maxwell's demon also dramatized the possibility of escaping from the dreadful inevitability of entropy increase. As I explained in the previous chapter, increasing entropy implies increasing disorder, a running down, a using up.

If entropy must constantly and continuously increase, then the universe is remorselessly running down, thus setting a limit (a long one, to be sure) on the existence of humanity. To some human beings, this ultimate end poses itself almost as a threat to their personal immortality, or as a denial of the omnipotence of God. There is, therefore, a strong emotional urge to deny that entropy *must* increase.

And in Maxwell's demon, they find substance for their denial. To be sure, the demon does not exist, but his essential attribute is his capacity to pick and choose among the moving molecules. Mankind's scientific ability is constantly increasing, and the day may come when he will be able, by some device, to duplicate the demon's function. Would he then not be able to decrease entropy?

Alas, there is a flaw in the argument. I hate to say this, but Maxwell cheated. The gas cannot be treated as an isolated system in the presence of the demon. The whole system would then consist of the gas *plus the demon.* In the process of selecting between fast and slow molecules, the demon's entropy would have to increase by an amount that more than made up for the decrease in entropy that he brings about in the gas.

Of course, I know that you suspect I have never really studied demons of any type, let alone one of the Maxwell variety. Nevertheless, I am confident of the truth of my statement, for the whole structure of scientific knowledge requires that the demon's entropy behave in this fashion.

And if man ever invents a device that will duplicate the activity of the demon, then you can bet that that device will undergo an entropy increase greater than the entropy decrease it will bring about. You will be perfectly safe to grant any odds at all.

The cold fact is that entropy increase cannot be beaten. No one has ever measured or demonstrated an overall entropy decrease anywhere in the universe under any circumstances.

But entropy is strictly applicable only to questions of energy flow. It can be defined in precise mathematical form in relation to heat and temperature and is capable of precise measurement where heat and temperature are concerned. What, then, if we depart from the field where entropy is applicable and carry the concept elsewhere? Entropy will then lose its rigorous nature and become a rather vague measure of orderliness or a rough indicator of the general nature of spontaneous change.

If we do that, can we work up an argument to demonstrate anything we can call an entropy decrease in the broad sense of the term?

Here's an example brought up by a friend of mine during an excellently heated evening of discourse. He said:

"As soon as we leave the world of energy, it is perfectly possible to decrease entropy. Men do it all the time. Here is *Webster's New International Dictionary*. It contains every word in *Hamlet* and *King Lear* in a particular order.

Shakespeare took those words, placed them in a different order and created the plays. Obviously, the words in the plays represent a much higher and more significant degree of order than do the words in the dictionary. Thus they represent, in a sense, a decrease in entropy. Where is the corresponding increase in entropy in Shakespeare? He ate no more, expended no more energy, than if he had spent the entire interval boozing at the Mermaid Tavern."

He had me there, I'm afraid, and I fell back upon a shrewd device I once invented as a particularly ingenious way out of such a dead end. I changed the subject.

But I returned to it in my thoughts at periodic intervals ever since. Since I feel (intuitively) that entropy increase is a universal necessity, it seemed to me I ought to be able to think up a line of argument that would make Shakespeare's creations of his plays an example of it.

And here's the way the matter now seems to me.

If we concentrate on the words themselves, then let's remember that Shakespeare's words make sense to us only because we understand English. If we knew only Polish, a passage of Shakespeare and a passage of the dictionary would be equally meaningless. Since Polish makes use of the Latin alphabet just as English does and since the letters are in the same order, it follows, however, that a Polish-speaking individual could find any English word in the dictionary without difficulty (even if he didn't know its meaning) and could find the same word in Shakespeare only by good fortune.

Therefore the words, considered only as words, are in more orderly form in the dictionary, and if the word order in Shakespeare is compared with the word order in the dictionary, the construction of the plays represents an increase in entropy.

But in concentrating on the words as literal objects (a subtle pun, by the way), I am, of course, missing the point. I do that only to remove the words themselves from the argument.

The glory of Shakespeare is not the physical form of the symbols he uses but the ideas and concepts behind those symbols. Let Shakespeare be translated into Polish and our Polish-speaking friend would far rather read Shakespeare than a Polish dictionary.

So let us forget words and pass on to ideas. If we do that, then it is foolish to compare Shakespeare to the dictionary. Shakespeare's profound grasp of the essence of humanity came not from any dictionary but from his observation and understanding of human beings.

If we are to try to detect the direction of entropy change, then, let us not compare Shakespeare's words to those in the dictionary, but Shakespeare's view of life to life itself.

Granted that no one in the history of human literature has so well interpreted the thoughts and emotions of mankind as well as Shakespeare has, it does not necessarily follow that he has improved on life itself.

It is simply impossible, in any cast of characters fewer than all men who have ever existed, in any set of passions weaker or less complex and intertangled than all that have ever existed, completely to duplicate life. Shakespeare has had to epitomize, and has done that superlatively well. In a cast of twenty and in the space of three hours, he exhibits more emotion and a more sensitive portrayal of various facets of humanity than any group of twenty real people could possibly manage in the interval of three real hours. In that respect he has produced what we might call a local decrease in entropy.

But if we take the entire system, and compare all of Shakespeare to all of life, surely it must be clear that Shakespeare has inevitably missed a vast amount of the complexity and profundity of the human mass and that his plays represent an overall increase of entropy.

And what is true for Shakespeare is true for all mankind's intellectual activity, it seems to me.

How I can best put this I am not certain, but I feel that nothing the mind of man can create is truly created out of nothing. All possible mathematical relationships; natural laws; combinations of words, lines, colors, sounds; all—everything—exists at least in potentiality. A particular man discovers one or another of these but does not create them in the ultimate sense of the word.

In seizing the potentiality and putting it into the concrete, there is always the possibility that something is lost in the translation, so to speak, and that represents an entropy increase.

Perhaps very little is lost, as for instance in mathematics. The relationship expressed by the Pythagorean theorem existed before Pythagoras, mankind, and the earth. Once grasped, it was grasped as it was. I don't see what can have been significantly lost in the translation. The entropy increase is virtually zero.

In the theories of the physical sciences, there is clearly less perfection and therefore a perceptible entropy increase. And in literature and the fine arts, intended to move our emotions and display us to ourselves, the entropy increase —even in the case of transcendent geniuses such as Sophocles and Beethoven—must be vast.

And certainly there is never an improvement on the potentiality; there is never a creation of that which has no potential existence. Which is a way of saying there is never a decrease in entropy.

I could almost wish, at this point, that I were in the habit of expressing myself in theological terms, for if I were, I might be able to compress my entire thesis into a sentence.

All knowledge of every variety (I might say) is in the mind of God—and the human intellect, even the best, in trying to pluck it forth can but "see through a glass, darkly."

Another example of what appears to be steadily decreasing entropy on a grand scale lies in the evolution of living organisms.

I don't mean by this the fact that organisms build up complex compounds from simple ones or that they grow and proliferate. This is done at the expense of solar energy, and it is no trick at all to show that an overall entropy increase is involved.

There is a somewhat more subtle point to be made. The specific characteristics of living cells (and therefore of living multicellular organisms, too, by way of the sex cells) are passed on from generation to generation by duplication of genes. These genes are immensely complicated compounds and, ideally, the duplication should be perfect.

But where are ideals fulfilled in this imperfect universe of ours? Errors will slip in, and these departures from perfection in duplication are called mutations. Since the errors are random and since there are many more ways in which a very complex chemical can lose complexity rather than gain it, the large majority of mutations are for the worse, in the sense that the cell or organism loses a capacity that its parent possessed.

(By analogy, there are many more ways in which a hard

The Granger Collection

WILLIAM SHAKESPEARE

In this article I mentioned Shakespeare as representing the kind of entropy reversal that science cannot deal with easily. The reason I did, is that I think Shakespeare is the greatest writer who ever lived. This is no novelty; lots of people think so; and you might suppose that I was only getting on the bandwagon to show how erudite and cultured I was. However, since I wrote a two-volume *Asimov's Guide to Shakespeare* (Doubleday, 1970), I can prove my interest to be legitimate.

In the introduction to that book, I said:

"Indeed, so important are Shakespeare's works that only the Bible can compare with them in their influence upon our language and thought. Shakespeare has said so many things so supremely well that we are forever finding ourselves thinking in his terms . . .

"I have a feeling that Shakespeare has even acted as a brake on the development of English . . . after three and a half centuries, Shakespeare's plays can be read quite easily and with only an occasional archaic word or phrase requiring translation. It is almost as though the English language dare not change so much as to render Shakespeare incomprehensible. That would be an unacceptable price to pay for change.

"In this respect, Shakespeare is even more important than the Bible. The King James version of the Bible is, of course, only a translation, although a supremely great one. If it becomes archaic there is nothing to prevent newer translations into more modern English. Indeed, such newer translations exist.

"How, though, can anyone ever dream of 'translating' Shakespeare into 'modern English'?"

Shakespeare, then, has reduced the entropy of the entire language in that his perfect phrases have entered it; and has reduced the entropy of our own thinking as well, in that we use phrases as tools—but, of course, that is not the kind of entropy this article deals with.

jar is likely to damage the workings of a delicate watch than to improve them. For that reason do not hit a stopped watch with a hammer and expect it to start again.)

This mutation-for-the-worse is in accord with the notion of increasing entropy. From generation to generation, the original gene pattern fuzzes out. There is an increase of disorder, each new organism loses something in the translation, and life degenerates to death. This should inevitably happen if only mutations are involved.

Yet this does not happen.

Not only does it not happen, but the reverse *does* happen.

On the whole, living organisms have grown more complex and more specialized over the aeons. Out of unicellular creatures came multicellular ones. Out of two germ layers came three. Out of a two-chambered heart came a four-chambered one.

This form of apparent entropy-decrease cannot be explained by bringing in solar energy. To be sure, an input of energy in reasonable amounts (short of the lethal level, that is) will increase the mutation rate. But it will not change the ratio of unfavorable to favorable changes. Energy input would simply drive life into genetic chaos all the faster.

The only possible way out is to have recourse to a demon (after the fashion of Maxwell) which is capable of picking and choosing among mutations, allowing some to pass and others not.

There is such a demon in actual fact, though, as far as I know, I am the only one who has called it that and drawn the analogy with Maxwell's demon. The English naturalist Charles Robert Darwin discovered the demon, so we can call it "Darwin's demon" even though Darwin himself called it "natural selection."

Those mutations which render a creature less fit to compete with other organisms for food, for mating or for self-defense, are likely to cause that creature to come to an untimely end. Those mutations which improve the creature's competing ability are likely to cause that creature to flourish. And, to be sure, fitness or lack of it relates only to the particular environment in which the creature finds itself. The best fins in the world would do a camel no good.

The effect of mutation *in the presence of natural selection*, then, is to improve continually the adjustment of a particular creature to its particular environment; and that is the direction of increasing entropy.

This may sound like arbitrarily defining entropy increase as the opposite of what it is usually taken to be—allowing entropy increase to signify increased order rather than increased disorder. This, however, is not so. I will explain by analogy.

Suppose you had a number of small figurines of various shapes and sizes lined up in orderly rank and file in the center of a large tray. If you shake the tray, the figurines will move out of place and become steadily more disordered.

This is analogous to the process of mutation without natural selection. Entropy obviously increases.

But suppose that the bottom of the tray possessed depressions into which the various figurines would just fit. If the figurines were placed higgledy-piggledy on the tray with not one figurine within a matching depression, then shaking the tray would allow each figurine to find its own niche and settle down into it.

Once a figurine found its niche through random motion, it would take a hard shake to throw it out.

This is analogous to the process of mutation with natural selection. Here entropy increases, for each figurine would have found a position where its center of gravity is lower than it would be in any other nearby position. And lowering the center of gravity is a common method of increasing entropy as, for instance, when a stone rolls downhill.

The organisms with which we are best acquainted have improved their fit to their environment by an increase in complexity in certain particularly noticeable respects. Consequently, we commonly think of evolution as necessarily proceeding from the simple to the complex.

This is an illusion. Where a simplifying change improves the fit of an organism to its environment, there the direction of evolution is from the complex to the simple. Cave creatures who live in utter darkness usually lose their eyes, although allied species living in the open retain theirs.

The reptiles went to a lot of trouble (so to speak) to develop two pairs of legs strong enough to lift the body clear of the ground. The snakes gave up those legs, slither on abdominal scales, and are the most successful of the contemporary reptiles.

Parasites undergo particularly great simplifications. A tapeworm suits itself perfectly to its environment by giving up the digestive system it no longer needs, the locomotor functions it doesn't use. It becomes merely an absorbing surface with a hooked proboscis with which to catch hold of the intestinal lining of its host, and the capacity to produce eggs and eggs and eggs and . . .

Such changes are usually called (with more than a faint air of disapproval) "degenerative." That, however, is only our prejudice. Why should we approve of some adjustments and disapprove of others? To the cold and random world of evolution, an adjustment is an adjustment.

If we sink to the biochemical level, then the human being has lost a great many synthetic abilities possessed by other species and, in particular, by plants and microorganisms. Our loss of ability to manufacture a variety of vitamins makes us dependent on our diet and, therefore, on the greater synthetic versatility of other creatures. This is as much a "degenerative" change as the tapeworm's abandonment of a stomach it no longer needs, but since we are prejudiced in our own favor, we don't mention it.

And, of course, no adjustment is final. If the environment changes; if the planetary climate becomes markedly colder, drier or damper; if a predator improves its efficiency or a new predator comes upon the scene; if a parasitic organism increases in infectivity or virulence; if the food supply dwindles for any reason—then an adjustment that was a satisfactory one before becomes an unsatisfactory one now and the species dies out.

The better the fit to a particular environment, the smaller the change in environment required to bring about extinction. Long-lived species are therefore those which pick a particularly stable environment; or those that remain somewhat generalized, being fitted well enough to one environment to compete successfully within it, but not so well as to be unable to shift to an allied environment if the first fails them.

In the case of Darwin's demon (as in that of Maxwell's demon), the question as to the role of human intelligence arises. Here it is not a matter of imitating the demon, but, rather, of stultifying it.

Many feel that the advance of human technology hampers the working of natural selection. It allows people with bad eyes to get along by means of glasses, diabetics to get along by means of insulin injections, the feeble-minded to get along by means of welfare agencies, and so on.

Some people call this "degenerative mutation pressure" and, as you can see from the very expression used, are concerned about it. Everyone without exception, as far as I know, considers this a danger to humanity, even though practically nobody proposes any non-humane solutions.

And yet is it necessarily a danger to humanity?

Let's turn degenerative mutation pressure upside down and see if it can't be viewed as something other than a danger.

In the first place, we can't really stultify Darwin's demon, for natural selection must work at all times, by definition. *Man is part of nature* and his influence is as much a natural one as is that of wind and water.

So let us assume that natural selection is working and ask what it is doing. Since it is fitting man to his environment (the only thing Darwin's demon can or does do), we must inquire as to what man's environment is. In a sense, it is all the world, from steaming rain jungle to frozen glacier, and the reason for that is that all contemporary men, however primitive, band together into societies that can more or less change the environment to suit their needs even if only by building a campfire or chipping a rock or tearing off a tree branch.

Consequently, it seems clear that the most important part of a man's environment is other men; or, if you prefer, human society. The vast majority of mankind, in fact, live as part of very complex societies that penetrate every facet of their lives.

If nearsightedness is not the handicap in New York that it would have been in a primitive hunting society, or if diabetes is not the handicap in Moscow that it would have been in a non-biochemical society, then why should there be any evolutionary pressures in favor of keeping unnecessarily good eyes and functional pancreases?

Man is to an increasing extent a parasite on human society; perhaps what we call "degenerative mutation pressure" is simply better fitting him to his new role, just as it better fit the tapeworm to its role. We may not like it, but it is a reasonable evolutionary change.

There are many among us who chafe at the restrictions of the crowded anthills we call cities, at the slavery to the clock hand, at the pressures and tensions. Some revolt by turning to delinquency, to "antisocial behavior." Others search out the dwindling areas where man can carry on a pioneer existence.

But if our anthills are to survive, we need those who will bend to its needs, who will avoid walking on grass, beating red lights and littering sidewalks. It is just the metabolically handicapped that can be relied on to do this, for they cannot afford to fight a society on which they depend, very literally, for life. A diabctic won't long for the great outdoors if it means his insulin supply will vanish.

If this is so, then Darwin's demon is only doing what comes naturally.

But of all environments, that produced by man's complex technology is perhaps the most unstable and rickety. In its present form, our society is not two centuries old, and a few nuclear bombs will do it in.

To be sure, evolution works over long periods of time and two centuries is far from sufficient to breed Homo technikos.

The closer this is approached, however, the more dangerour would become any shaking of our social structure. The destruction of our technological society in a fit of nuclear peevishness would become disastrous even if there were many millions of immediate survivors.

The environment toward which they were fitted would be gone, and Darwin's demon would wipe them out remorselessly and without a backward glance.

Fourteen
A PIECE OF THE ACTION

When my book *I, Robot* was reissued by the estimable gentlemen of Doubleday & Company, it was with a great deal of satisfaction that I noted certain reviewers (possessing obvious intelligence and good taste) beginning to refer to it as a "classic."

"Classic" is derived in exactly the same way, and has precisely the same meaning, as our own "first-class" and our colloquial "classy"; and any of these words represents my own opinion of *I, Robot*, too; except that (owing to my modesty) I would rather die than admit it. I mention it here only because I am speaking confidentially.

However, "classic" has a secondary meaning that displeases me. The word came into its own when the literary men of the Renaissance used it to refer to those works of the ancient Greeks and Romans on which they were modeling their own efforts. Consequently, "classic" has come to mean not only *good*, but also *old*.

Now *I, Robot* first appeared a number of years ago and some of the material in it was written . . . Well, never mind. The point is that I have decided to feel a little hurt at being considered old enough to have written a classic, and therefore I will devote this chapter to the one field where "classic" is rather a term of insult.

Naturally, that field must be one where to be old is, almost automatically, to be wrong and incomplete. One may talk about Modern Art or Modern Literature or Modern Furniture and sneer as one speaks, comparing each, to their disadvantage, with the greater work of earlier ages. When one speaks of Modern Science, however, one removes one's hat and places it reverently upon the breast.

In physics, particularly, this is the case. There is Modern Physics and there is (with an offhand, patronizing half-smile) Classical Physics. To put it into Modern Terminology, Modern Physics is in, man, in, and Classical Physics is like squaresville.

What's more, the division in physics is sharp. Everything after 1900 is Modern; everything before 1900 is Classical.

That looks arbitrary, I admit; a strictly parochial twentieth-century outlook. Oddly enough, though, it is perfectly legitimate. The year 1900 saw a major physical theory entered into the books and nothing has been quite the same since.

By now you have guessed that I am going to tell you about it.

The problem began with German physicist Gustav Robert Kirchhoff who, with Robert Wilhelm Bunsen (popularizer of the Bunsen burner), pioneered in the development of spectroscopy in 1859. Kirchhoff discovered that each element, when brought to incandescence, gave off certain characteristic frequencies of light; and that the vapor of that element, exposed to radiation from a source hotter than itself, absorbed just those frequencies it itself emitted when radiating. In short, a material will absorb those frequencies which, under other conditions, it will radiate; and will radiate those frequencies which, under other conditions, it will absorb.

But suppose that we consider a body which will absorb all frequencies of radiation that fall upon it—absorb them completely. It will then reflect none and will therefore appear absolutely black. It is a "black body." Kirchhoff pointed out that such a body, if heated to incandescence, would then necessarily have to radiate all frequencies of radiation. Radiation over a complete range in this manner would be "black-body radiation."

Of course, no body was absolutely black. In the 1890s, however, a German physicist named Wilhelm Wien thought of a rather interesting dodge to get around that. Suppose you had a furnace with a small opening. Any radiation that passes through the opening is either absorbed by the rough wall opposite or reflected. The reflected radiation strikes another wall and is again partially absorbed. What is reflected strikes another wall, and so on. Virtually none of the radiation survives to find its way out the small opening again. That small opening, then, absorbs the radiation and, in a manner of speaking, reflects none. It is a black body. If the furnace is heated, the radiation that streams out of that small opening should be black-body radiation and should, by Kirchhoff's reasoning, contain all frequencies.

Wien proceeded to study the characteristics of this black-body radiation. He found that at any temperature, a wide spread of frequencies was indeed included, but the spread was not an even one. There was a peak in the middle. Some intermediate frequency was radiated to a greater extent than other frequencies either higher or lower than that peak frequency. Moreover, as the temperature was increased, this peak was found to move toward the higher frequencies. If the absolute temperature were doubled, the frequency at the peak would also double.

But now the question arose: *Why* did black-body radiation distribute itself like this?

To see why the question was puzzling, let's consider infrared light, visible light, and ultraviolet light. The frequency range of infrared light, to begin with, is from one hundred billion (100,000,000,000) waves per second to four hundred trillion (400,000,000,000,000) waves per second. In order to make the numbers easier to handle, let's divide by a hundred billion and number the frequency not in individual waves per second but in hundred-billion-wave packets per second. In that case the range of infrared would be from 1 to 4000.

Continuing to use this system, the range of visible light would be from 4000 to 8000; and the range of ultraviolet light would be from 8000 to 300,000.

Now it might be supposed that if a black body absorbed all radiation with equal ease, it ought to give off all radiation with equal ease. Whatever its temperature, the energy it had to radiate might be radiated at any frequency, the particular choice of frequency being purely random.

But suppose you were choosing numbers, *any* numbers with honest randomness, from 1 to 300,000. If you did this repeatedly, trillions of times, 1.3 per cent of your numbers would be less than 4000; another 1.3 per cent would be between 4000 and 8000, and 97.4 per cent would be between 8000 and 300,000.

This is like saying that a black body ought to radiate 1.3 per cent of its energy in the infrared, 1.3 per cent in visible light, and 97.4 per cent in the ultraviolet. If the temperature went up and it had more energy to radiate, it ought to radiate more at every frequency but the relative amounts in each range ought to be unchanged.

And this is only if we confine ourselves to nothing of still higher frequency than ultraviolet. If we include the x-ray frequencies, it would turn out that just about nothing should come off in the visible light at any temperature. Everything would be in ultraviolet and x-rays.

An English physicist, Lord Rayleigh, worked out an equation which showed exactly this. The radiation emitted by a black body increased steadily as one went up the frequencies. However, in actual practice, a frequency peak was reached after which, at higher frequencies still, the quantity of radiation decreased again. Rayleigh's equation was interesting but did not reflect reality.

Physicists referred to this prediction of the Rayleigh equation as the "Violet Catastrophe"—the fact that every body that had energy to radiate ought to radiate practically all of it in the ultraviolet and beyond.

Yet the whole point is that the Violet Catastrophe does not take place. A radiating body concentrated its radiation in the low frequencies. It radiated chiefly in the infrared at temperatures below, say, 1000° C., and radiated mainly in the visible region even at a temperature as high as 6000° C., the temperature of the solar surface.

Yet Rayleigh's equation was worked out according to the very best principles available anywhere in physical theory—at the time. His work was an ornament of what we now call Classical Physics.

Wien himself worked out an equation which described the frequency distribution of black-body radiation in the high-frequency range, but he had no explanation for why it worked there, and besides it only worked for the high-frequency range, not for the low-frequency.

Black, black, black was the color of the physics mood all through the later 1890s.

But then arose in 1899 a champion, a German physicist, Max Karl Ernst Ludwig Planck. He reasoned as follows . . .

If beautiful equations worked out by impeccable reasoning from highly respected physical foundations do not describe the truth as we observe it, *then* either the reasoning or the physical foundations or both are wrong.

And *if* there is nothing wrong about the reasoning (and nothing wrong could be found in it), *then* the physical foundations had to be altered.

The physics of the day required that all frequencies of

light be radiated with equal probability by a black body, and Planck therefore proposed that, on the contrary, they were *not* radiated with equal probability. Since the equal-probability assumption required that more and more light of higher and higher frequency be radiated, whereas the reverse was observed, Planck further proposed that the probability of radiation ought to decrease as frequency increased.

In that case, we would now have two effects. The first effect would be a tendency toward randomness which would favor high frequencies and increase radiation as frequency was increased. Second, there was the new Planck effect of decreasing probability of radiation as frequency went up. This would favor low frequencies and decrease radiation as frequency was increased.

In the low-frequency range the first effect is dominant, but in the high-frequency range the second effect increasingly overpowers the first. Therefore, in black-body radiation, as one goes up the frequencies, the amount of radiation first increases, reaches a peak, then decreases again—exactly as is observed.

Next, suppose the temperature is raised. The first effect can't be changed, for randomness is randomness. But suppose that as the temperature is raised, the probability of emitting high-frequency radiation increases. The second effect, then, is steadily weakened as the temperature goes up. In that case, the radiation continues to increase with increasing frequency for a longer and longer time before it is overtaken and repressed by the gradually weakening second effect. The peak radiation, consequently, moves into higher and higher frequencies as the temperature goes up—precisely as Wien had discovered.

On this basis, Planck was able to work out an equation that described black-body radiation very nicely both in the low-frequency and high-frequency range.

However, it is all very well to say that the higher the frequency the lower the probability of radiation, by *why?* There was nothing in the physics of the time to explain that, and Planck had to make up something new.

Suppose that energy did not flow continuously, as physicists had always assumed, but was given off in pieces. Suppose there were "energy atoms" and these increased in size as frequency went up. Suppose, still further, that light of a particular frequency could not be emitted unless enough energy had been accumulated to make up an "energy atom" of the size required by that frequency.

The higher the frequency the larger the "energy atom" and the smaller the probability of its accumulation at any given instant of time. Most of the energy would be lost as radiation of lower frequency, where the "energy atoms" were smaller and more easily accumulated. For that reason, an object at a temperature of 400° C. would radiate its heat in the infrared entirely. So few "energy atoms" of visible light size would be accumulated that no visible glow would be produced.

As temperature went up, more energy would be generally available and the probabilities of accumulating a high-frequency "energy atom" would increase. At 6000° C. most of the radiation would be in "energy atoms" of visible light, but the still larger "energy atoms" of ultraviolet would continue to be formed only to a minor extent.

But how big is an "energy atom"? How much energy does it contain? Since this "how much" is a key question, Planck, with admirable directness, named the "energy atom" a *quantum*, which is Latin for "how much?" The plural is *quanta*.

For Planck's equation for the distribution of black-body radiation to work, the size of the quantum had to be directly proportional to the frequency of the radiation. To express this mathematically, let us represent the size of the quantum, or the amount of energy it contains, by e (for energy). The frequency of radiation is invariably represented by physicists by means of the Greek letter *nu* (ν).

If energy (e) is proportional to frequency (ν), then e must be equal to ν multiplied by some constant. This constant, called *Planck's constant*, is invariably represented as h. The equation, giving the size of a quantum for a particular frequency of radiation, becomes:

$$e = h\nu \qquad \text{(Equation 23)}$$

It is this equation, presented to the world in 1900, which is the Continental Divide that separates Classical Physics from Modern Physics. In Classical Physics, energy was considered continuous; in Modern Physics it is considered to be composed of quanta. To put it another way, in Classical

Physics the value of *h* is considered to be 0; in Modern Physics it is considered to be greater than 0.

It is as though there were a sudden change from considering motion as taking place in a smooth glide, to motion as taking place in a series of steps.

There would be no confusion if steps were long galumphing strides. It would be easy, in that case, to distinguish steps from a glide. But suppose one minced along in microscopic little tippy-steps, each taking a tiny fraction of a second. A careless glance could not distinguish that from a glide. Only a painstaking study would show that your head was bobbing slightly with each step. The smaller the steps, the harder to detect the difference from a glide.

In the same way, everything would depend on just how big individual quanta were; on how "grainy" energy was. The size of the quanta depends on the size of Planck's constant, so let's consider that for a while.

If we solve Equation 23 for *h*, we get:

$$h = e/\nu \qquad \text{(Equation 24)}$$

Energy is very frequently measured in ergs (see Chapter 9). Frequency is measured as "so many per second" and its units are therefore "reciprocal seconds" or "1/second."

We must treat the units of *h* as we treat *h* itself. We get *h* by dividing *e* by ν; so we must get the units of *h* by dividing the units of *e* by the units of ν. When we divide ergs by 1/second we are multiplying ergs by seconds, and we find the units of *h* to be "erg-seconds." A unit which is the result of multiplying energy by time is said, by physicists, to be one of "action." Therefore, Planck's constant is expressed in units of action.

Since the nature of the universe depends on the size of Planck's constant, we are all dependent on the size of the piece of action it represents. Planck, in other words, had sought and found *the* piece of the action. (I understand that others have been searching for a piece of the action ever since, but where's the point since Planck has found it?)

And what is the exact size of *h?* Planck found it had to be very small indeed. The best value, currently accepted, is: 0.0000000000000000000000000066256 erg-seconds or 6.6256×10^{-27} erg-seconds.

Now let's see if I can find a way of expressing just how

The Granger Collection

MAX KARL ERNST LUDWIG PLANCK

Max Planck was born in Kiel, Germany, on April 23, 1858, and passed his fortieth birthday without having done anything worthy of special remark, though he was, of course, a perfectly adequate physicist.

In 1900, however, he worked out the revolutionary quantum theory, a towering achievement which extended and improved the basic concepts of physics. It was so revolutionary, in fact, that almost no physicist, including Planck himself, could bring himself to accept it. (Planck later said that the only way a revolutionary theory could be accepted was to wait until all the old scientists had died.) It was Einstein's use of quantum theory in connection with the photoelectric effect, in 1905, that really established it, and in 1918, Planck received a Nobel Prize in physics for it.

In 1930, Planck became president of the Kaiser Wilhelm Society of Berlin, and it was renamed the Max Planck Society. His old age saw his renown in the world of science second only to that of Einstein. Nor was he too old to resist Hitler firmly in the days of Nazi ascendancy. He conceived it his duty to remain in Germany, but never lent his name and prestige to the regime. He intervened personally (but unsuccessfully) with Hitler on behalf of his Jewish colleagues and was forced to resign his post in 1937.

During World War II, his older son was killed in action, and his younger was executed for taking part in a plot against Hitler. Planck survived the war, however, and lived to see Nazism destroyed. He was rescued by American forces in 1945, and was then renamed president of the Max Planck Society. He died in Göttingen, on October 3, 1947, in his ninetieth year.

small this is. The human body, on an average day, consumes and expends about 2500 kilocalories in maintaining itself and performing its tasks. One kilocalorie is equal to 1000 calories, so the daily supply is 2,500,000 calories.

One calorie, then, is a small quantity of energy from the human standpoint. It is $\frac{1}{2,500,000}$ of your daily store. It is the amount of energy contained in $\frac{1}{113,000}$ of an ounce of sugar, and so on.

Now imagine you are faced with a book weighing one pound and wish to lift it from the floor to the top of a bookcase three feet from the ground. The energy expended in lifting one pound through a distance of three feet against gravity is just about 1 calorie.

Suppose that Planck's constant were of the order of a calorie-second in size. The universe would be a very strange place indeed. If you tried to lift the book, you would have to wait until enough energy had been accumulated to make up the tremendously sized quanta made necessary by so large a piece of action. Then, once it was accumulated, the book would suddenly be three feet in the air.

But a calorie-second is equal to 41,850,000 erg-seconds, and since Planck's constant is such a minute fraction of one erg-second, a single calorie-second equals 6,385,400,000,000,000,000,000,000,000,000,000 Planck's constants, or 6.3854×10^{33} Planck's constants, or about six and a third decillion Planck's constants. However you slice it, a calorie-second is equal to a tremendous number of Planck's constants.

Consequently, in any action such as the lifting of a one-pound book, matters are carried through in so many trillions of trillions of steps, each one so tiny, that motion seems a continuous glide.

When Planck first introduced his "quantum theory" in 1900, it caused remarkably little stir, for the quanta seemed to be pulled out of midair. Even Planck himself was dubious—not over his equation describing the distribution of black-body radiation, to be sure, for that worked well; but about the quanta he had introduced to explain the equation.

Then came 1905, and in that year a 26-year-old theoretical physicist, Albert Einstein, published five separate scientific papers on three subjects, any one of which would have been enough to establish him as a first-magnitude star in the scientific heavens.

In two, he worked out the theoretical basis for "Brownian motion" and, incidentally, produced the machinery by which the actual size of atoms could be established for the first time. It was one of these papers that earned him his Ph.D.

In the third paper, he dealt with the "photoelectric effect" and showed that although Classical Physics could not explain it, Planck's quantum theory could.

This really startled physicists. Planck had invented quanta merely to account for black-body radiation, and here it turned out to explain the photoelectric effect, too, something entirely different. For quanta to strike in two different places like this, it seemed suddenly very reasonable to suppose that they (or something very like them) actually existed.

(Einstein's fourth and fifth papers set up a new view of the universe which we call "The Special Theory of Relativity." It is in these papers that he introduced his famous equation $e = mc^2$; see Chapter 9.

These papers on relativity, expanded into a "General Theory" in 1915, are the achievements for which Einstein is known to people outside the world of physics. Just the same, in 1921, when he was awarded the Nobel Prize for Physics, it was for his work on the photoelectric effect and *not* for his theory of relativity.)

The value of h is so incredibly small that in the ordinary world we can ignore it. The ordinary gross events of everyday life can be considered as though energy were a continuum. This is a good "first approximation."

However, as we deal with smaller and smaller energy changes, the quantum steps by which those changes must take place become larger and larger in comparison. Thus, a flight of stairs consisting of treads 1 millimeter high and 3 millimeters deep would seem merely a slightly roughened ramp to a six-foot man. To a man the size of an ant, however, the steps would seem respectable individual obstacles to be clambered over with difficulty. And to a man the size of a bacterium, they would be mountainous precipices.

In the same way, by the time we descend into the world within the atom the quantum step has become a gigantic thing. Atomic physics cannot, therefore, be described in Classical terms, not even as an approximation.

The first to realize this clearly was the Danish physicist Niels Bohr. In 1913 Bohr pointed out that if an electron absorbed energy, it had to absorb it a whole quantum at a time and that to an electron a quantum was a large piece of energy that forced it to change its relationship to the rest of the atom drastically and all at once.

Bohr pictured the electron as circling the atomic nucleus in a fixed orbit. When it absorbed a quantum of energy, it suddenly found itself in an orbit farther from the nucleus—there was no in-between, it was a one-step proposition.

Since only certain orbits were possible, according to Bohr's treatment of the subject, only quanta of certain size could be absorbed by the atom—only quanta large enough to raise an electron from one permissible orbit to another. When the electrons dropped back down the line of permissible orbits, they emitted radiations in quanta. They emitted just those frequencies which went along with the size of quanta they could emit in going from one orbit to another.

In this way, the science of spectroscopy was rationalized. Men understood a little more deeply why each element (consisting of one type of atom with one type of energy relationships among the electrons making up that type of atom) should radiate certain frequencies, and certain frequencies only, when incandescent. They also understood why a substance that could absorb certain frequencies should also emit those same frequencies under other circumstances.

In other words, Kirchhoff had started the whole problem and now it had come around full-circle to place his empirical discoveries on a rational basis.

Bohr's initial picture was oversimple; but he and other men gradually made it more complicated, and capable of explaining finer and finer points of observation. Finally, in 1926, the Austrian physicist Erwin Schrödinger worked out a mathematical treatment that was adequate to analyze the workings of the particles making up the interior of the atom according to the principles of the quantum theory. This was called "quantum mechanics," as opposed to the "classical mechanics" based on Newton's three laws of motion and it is quantum mechanics that is the foundation of Modern Physics.

Fifteen

THE CERTAINTY OF UNCERTAINTY

In high school, one of the pieces of literature I was required to read was James Barrie's *The Admirable Crichton*. I reacted to it quite emotionally, but that is not the point at this moment. What is the point is that one of the characters, a wellborn young goof named Ernest, had carefully polished up an epigram which he sprang several times during the play.

It went, "After all, I'm not young enough to know everything."

And whenever he said it, someone would answer (impatiently, if the head of the family; wearily, if one of the ladies; paternally, if the competent butler), "You mean you're not *old* enough to know everything."

Ernest nearly died of frustration and so did I, for *I* knew what he meant.*

The memory of the epigram stays with me because, as it happens, nineteenth-century science was young enough to know everything. Near the opening of that century, the French astronomer Pierre Simon de Laplace had said, "If we knew the exact position and velocity of every particle in the universe at any one particular moment, then we could work out all the past and future of the universe."

The universe, in other words, was completely determinate, and when I read of this (being a convinced determinist myself) I fairly licked my lips with pleasure.

Of course, I understood that we didn't actually know the exact position and velocity of every particle in the universe at any particular moment, and that we almost certainly never would. However, we could know them *in principle* and that made the universe completely determinate *in principle*.

* He meant that young people *thought* they knew everything, but that as they grew older and wiser they realized they didn't. Good Lord!

Wasn't it a great feeling, though, to be young enough to know everything!

But alas we grow older and wiser and knowledge slips through our fingers after all and leaves us naked in a cold and hostile universe. My day of reckoning came in 1936 when I read *Uncertainty*, a two-part serial by John W. Campbell, Jr., in *Amazing Stories*. For the first time in my life I found out that the universe was not completely determinate, and couldn't be completely determinate *even in principle*.

So let's talk about uncertainty.

The basic principle is this: The very act of measurement alters the quantity being measured.

The most common example used to illustrate this is the measurement of the temperature of a container of hot water. The easiest way is to insert a thermometer, but if the thermometer is at room temperature, as it probably is, it withdraws heat from the water and when it finally registers a temperature, that temperature is slightly less than the temperature was before the thermometer was inserted.

This difficulty might be circumvented if the thermometer happened to be exactly at the temperature of the water to begin with. But in that case, how would you know the right temperature to begin with unless you measured it first?

Of course, the thermometer might just happen to be at the right temperature, and you could tell that because after insertion into the water the reading would remain the same as before. The thermometer would neither gain nor lose heat; and the temperature of the water would remain as it was and you would have the true and exact temperature.

You wouldn't even have to rely on pure chance. You could, for instance, perform a "thought experiment" (that is one which is conceivable, but which supposes conditions far too ideal and tedious to put into actual practice). We could divide our sample of water into independent sections each at the same temperature as all the rest.

We can then stick thermometers into the various sections, each thermometer being carefully warmed in advance to a different temperature, at one-degree intervals. One of those thermometers will register the same temperature before and after and there you would have the true and exact temperature.

Well, true and exact to the nearest degree anyway. Of course, that is just a detail, we could work with thermometers adjusted to differences of tenths of a degree, or hundredths, or thousandths. (In a thought experiment there is scarcely any limit to the refinement of our instruments.) But then there would always be a next stage in refinement.

Another way of making things finer is to use a smaller and smaller thermometer. The smaller the thermometer, the less heat it will be able to take up or give off and the smaller the deviation from the truth that it will produce. By taking measurements with thermometers at different size, one might even be able to calculate what the temperature would be with a "zero-size" thermometer.

Of course, though, to make a true, ultimate calculation of the temperature with a zero-size thermometer, you must be able to read the temperatures given by the various finite-size thermometers with infinite exactness, and you can't do that.

In short, for various reasons, a perfectly exact measurement cannot be made, and there will always be a residual uncertainty, however small.

Of course, you can shrug this off as merely a philosophic point of no practical importance. You may not be able to make a measurement infinitely exact, but you can make it as exact as necessary. If the necessity for refinement sharpens, you need then only make sharper measurements. The uncertainty of your measurement will never be zero but (the old argument went) it can be made to approximate zero as closely as you wish.

This, however, is true only if it is taken for granted that you can make the effect on the measurement of the act of measurement itself very small. For this, the measuring device must be very small, or, at least, contain a very small component. But what if there is an ultimate limit to smallness and, if you try to measure some property of that ultimately small object, you must make use of a measuring device as big as itself, or bigger.

Then, too, suppose that in measuring one property of a system, you upset a second property; and that the closer and more accurate the measurement of the first, the more wildly disturbed is the second. To gain certainty in one place at the cost of gross uncertainty in another is no true gain.

Consider, for instance, the electron, which has a mass of 9.1×10^{-28} grams.† This is, as far as we know, a rock-bottom minimum to mass. No object that possesses mass at all possesses less mass than the electron.

Suppose, then, we want to measure some of the properties of an electron speeding by. With Laplace's great statement in mind, we want to determine the position and velocity of that electron at some given moment. It we do that there will still remain an enormous step to the ultimate goal of learning the position and velocity of *all* particles at one particular moment, but the longest journey begins with but a single step. Let's concentrate on one electron to begin with.

The usual method of determining the position of anything is to receive light radiated by it, or to strike it with light and receive the reflection. In short, we see the object, and therefore know where it is.

An ordinary object is not appreciably affected by the light it reflects, but an electron is so small that it could be strenuously affected by that light. The idea, therefore, would be to use a very faint beam of light, one so faint that the electron would not be appreciably affected.

Unfortunately, there is a limit to light's faintness. Just as mass comes in thus-small-and-no-smaller units, so do all forms of energy. The least amount of light we can use is one photon, and if we try to send a photon of ordinary light at an electron, the wavelength associated with that photon is so long that it "steps over" the electron, and we can't see it.

We must use radiation of far shorter wavelength, an x ray or, better, a gamma ray, and receive the reflection by instruments. That is fine, but the shorter the wavelength, the greater the energy content of the photon. If a gamma-ray photon hits an electron, that electron might as well be kicked by a mule. It goes skittering off somewhere.

In other words, we may determine the position of the electron at a given moment, but the very act of that determination alters the velocity of that same electron at that same moment so that we cannot be sure what the velocity is.

Far more complicated thought experiments have been concocted and it turns out always that determining an elec-

† Or by the best current value, as I said in an earlier footnote, 9.109534×10^{-28} grams.

tron's position alters its velocity and that determining an electron's velocity alters its position. A *simultaneous* determination of both properties, with the uncertainty in *each* as close to zero as you wish, turns out to be impossible. At least, no one has ever devised a thought experiment which would yield such simultaneous exactness. Even Einstein tried and even Einstein failed.

In 1927, the German physicist Werner Heisenberg formalized this view by announcing what he called the "uncertainty principle." This is now accepted as one of the fundamental generalizations of the physical universe; as fundamental, universal, and inescapable as a generalization can be. In fact, insofar as there can be certainty about anything at all in the universe, there is the certainty of uncertainty.

Heisenberg expressed the principle in an equation which we can work out as follows. Let's symbolize position as p and momentum (which is equal to the mass of an object times its velocity) as mv. Uncertainty in a measurement is often expressed as the Greek capital "delta," which is just a triangle. The uncertainty in the measurement of position is therefore Δp and that in the measurement of momentum is Δmv. The equation expressing Heisenberg's uncertainty principle is therefore:

$$(\Delta p)\ (\Delta mv) = \frac{h}{2\pi} \qquad \text{(Equation 25)}$$

The symbol, h, is Planck's constant, and π (the Greek letter "pi") is the well-known ratio of the circumference of a circle to its diameter.

If we measure position in centimeters, mass in grams, and velocity in centimeters per second, then the approximate value of h comes out to be 6.6256×10^{-27} erg-seconds. The approximate value of π is, of course, 3.1416. We can therefore express Equation 25 (very nearly) as:

$$(\Delta p)\ (\Delta mv) = 10^{-27} \qquad \text{(Equation 26)}$$

In a way, uncertainty arises out of the "graininess" of the universe; out of the fact that energy and mass come in packages of fixed size; that size being ultimately determined by the value of Planck's constant. If Planck's constant were equal to zero, there would be no uncertainty at all. If Planck's constant were quite large, matters would grow so uncertain that the universe would seem chaotic.

The situation is analogous to that of a newspaper photograph built up out of black and white dots, or a television picture built up out of closely spaced lines. The coarser the dots or the lines, the fuzzier the picture and the poorer the detail; the finer the dots or the lines, the clearer the picture and the sharper the detail.

The graininess of the universe as represented by Planck's constant is fine indeed, exceedingly fine; so fine that before the twentieth century, the graininess was never noticed. It had seemed that all measurements could be made as accurately as time and patience would permit and that, in principle, this accuracy could be made of unlimited closeness to the ultimate of zero uncertainty.

The question now is whether the graininess of the universe is so fine that even now, in the twentieth century, it might not be fine enough to ignore; whether it might not be a point of philosophic interest only and of no concern to practical men, or even to practical scientists.

Let's consider Equation 26 again. Heisenberg spoke of the uncertainty of the measurement of momentum rather than of velocity because as the velocity of an object increases, so does its mass, and the two are naturally treated together. However, mass alters appreciably only at very high velocities, and if velocities are kept below, say, a thousand miles a second, we can, without too great an error, consider the value of *m* to be constant.

In that case we can deal with the uncertainty in velocity rather than momentum and write Equation 26 as:

$$(\Delta p) m (\Delta v) = 10^{-27} \qquad \text{(Equation 27)}$$

Transposing *m*, we have:

$$(\Delta p)\ (\Delta v) = \frac{10^{-27}}{m} \qquad \text{(Equation 28)}$$

Now we have an equation which tells us how to calculate the uncertainty in the simultaneous measurement of position and velocity of any particle, just the pair of measurements Laplace wanted to make. Under the circumstances outlined in Equation 28, we can see that we don't want to determine position *too* accurately for that will throw the measurement of velocity way off. Nor do we want to be too accurate about velocity at the expense of position. Let's make the healthy compromise of treating both position and velocity alike and making the measurements in such a way that the uncertainty in both cases is equal. We can do better with either one taken separately but we cannot possibly do better with both taken together.

Of the two measurements, that of position is the more dramatic. It is easy to see that we might not be certain of the exact velocity of an object, but surely (common sense tells us) we ought to know where it *is*, for heaven's sake. So, in Equation 28, let us let uncertainty in position equal uncertainty in velocity (numerically only, for the units will still differ) and let's represent both as (Δp). That gives us:

$$(\Delta p)^2 = \frac{10^{-27}}{m} \qquad \text{(Equation 29)}$$

$$\Delta p = \frac{3.2 \times 10^{-14}}{\sqrt{m}} \qquad \text{(Equation 30)}$$

We can put Equation 30 to work. We are dealing with measurements of mass in grams, so let's consider the uncertainty involved in measuring the position and velocity of a 1-gram mass. (This is not a large mass; 1 gram is equal to about 1/28 of an ounce.)

If m is set equal to 1, then the uncertainty in position comes out, according to Equation 30, to be 3.2×10^{-14} centimeters. Another way of putting this uncertainty in position is 0.000000000000032 centimeters.

Modern techniques do not make it possible to measure the position of a gram weight that closely and no one in his right mind would need to measure it that closely for any practical purpose. However, it is important to remember that no matter how we refine our measurements, no matter how much time we take, how much ingenuity we expend, it is impossible to measure the position of a gram weight to an uncertainty of less than 0.000000000000032 centimeters; at least not without introducing a larger uncertainty in velocity; and it is *both* position and velocity that Laplace requires.

The Bettmann Archive, Inc.

WERNER KARL HEISENBERG

Heisenberg, the son of a professor of Byzantine history, was born in Duisberg, Germany, on December 5, 1901. From his youth, he had right-wing sympathies and in the unsettled days after World War I, he was not too scholarly to engage in street fights against the Communists in Munich. In later years, he found outlets for his physical energies in becoming an enthusiastic mountain climber.

He worked under such scientists as Niels Bohr and Arnold Sommerfeld and got his Ph.D. in 1923. Bohr and Sommerfeld were trying to picture the inner structure of the atom and were thinking in terms of waves. Heisenberg decided to do neither. In 1927, he was on a vacation on a North Sea island, where he had gone to escape the discomfort of hay fever, and there, in relative isolation, he worked out methods for using "matrix mechanics" in connection with atomic structure. He represented the various atoms in terms of arrays of quantities which, properly manipulated, gave the wavelengths of the characteristic spectral lines of each particular atom. In this way, all attempts at visualization were done away with.

In the same year he announced the uncertainty principle and for this he obtained the Nobel Prize in physics in 1932.

Heisenberg was one of the few top-notch scientists who found themselves able to work under the Nazis. He even accepted high positions under them and during World War II was in charge of German research on the nuclear bomb. Before success could be achieved, the war came to an end. Heisenberg was director of the Max Planck Institute at Berlin, but after the war, he moved into West Germany and became director of the Max Planck Institute for Physics at Göttingen.

Yes, yes, you may say, but 0.0000000000000032 centimeters is close enough. If we can get all the particles in the universe that closely positioned and velocitied, we can still stretch things far back into the past and far forward into the future.

Ah, but this unavoidable uncertainty of 0.000000000000032 centimeters is for a 1-gram weight. If you look at Equation 30 you will see that as mass decreases the value of Δp must increase. For instance, in Table 7, I have listed the uncertainties for a variety of objects that are considerably smaller than 1 gram in mass.

As you see, the graininess of the universe seems to be sufficiently fine for us to remain undisturbed by uncertainty even in the case of ordinary microscopic objects. If we measure the position of a bacterium down to an uncertainty of merely three hundred-millionths of a centimeter, surely we can't complain.

Only when we penetrate lower than the merely microscopic and approach the atomic and subatomic are we really in trouble. Only then does the uncertainty principle become something that can't be shrugged off as merely academic.

TABLE 7—Some Uncertainties

OBJECT	APPROXIMATE MASS (GRAMS)	UNCERTAINTY OF POSITION (CENTIMETERS)
Ameba	4×10^{-6}	0.000000000016
Bacterium	1×10^{-12}	0.000000032
Gene	4×10^{-17}	0.000005
Uranium atom	4×10^{-22}	0.0016
Proton	1.6×10^{-24}	0.025
Electron	9.1×10^{-28}	1.1

In fact, the situation is even worse at that lower end of the scale than it appears to be in Table 7. You might console yourself by saying that even a proton can be pinned down to within a fortieth of a centimeter, which isn't so awful, and that only the electron gives us trouble.

However, why use an arbitrary, unchangeable unit of length such as the centimeter? Why not adjust the unit to the object, by taking the object's own diameter as the measure of position? The reason for this is simple. If you yourself shift your position by a hundredth of a centimeter, this is unimportant and the ordinary observer will neither know nor care that you have moved. If, however, an ameba shifts its position by a hundredth of a centimeter, it moves the distance of its own diameter and anyone observing the ameba under a microscope would see the move and find it highly significant. So let's prepare Table 8.

From this standpoint, affairs on the atomic and subatomic level are wildly uncertain, tremendously uncertain. If we try to ignore uncertainty on the atomic and subatomic level the results are simply grotesque. We cannot possibly view subatomic particles as tiny billiard balls because we can never pin down the position of such a tiny billiard ball. The best we can do (even if we are willing to increase the uncertainty in velocity tremendously) is to view it as a fuzzy object.

TABLE 8—More Uncertainties

OBJECT	APPROXIMATE DIAMETER (CENTIMETERS)	UNCERTAINTY OF POSITION (DIAMETERS)
Ameba	0.016	0.000000001
Bacterium	0.0001	0.0003
Gene	0.0000034	1.5
Uranium atom	0.00000001	160,000
Proton	0.0000000000001	250,000,000,000
Electron	0.0000000000001	11,000,000,000,000

You might also speak of a particle that exists but that you can't detect as a particle, and to suppose that this particle has a particular probability of being here, or there, or in the other place. That is why it is so useful to consider particles as possessing wave-properties. Not only does the wave take up room and appear "fuzzy" but the equations that describe the waves also describe the probabilities of the particle being at this or that point in space.

The graininess of the universe is so coarse, relatively, at the subatomic level, that there is no way in which we can get a meaningful picture of atomic structure by using analogies from the ordinary world, where the graininess of the universe appears so fine that it can be ignored altogether. The best that can be done (and what I always do, for instance) is to advance misleading simplifications and hope they don't mislead too badly.

Of course, if the universe is grainy, it would be interesting to find evidence of the grains on a large scale, too, and not just among protons and electrons.

We can, indeed, imagine large-scale situations where the principle of uncertainty would make itself manifest. One such situation is described in an excellent book entitled *The Laws of Physics* by Milton A. Rothman (Basic Books, 1963).

Imagine, says Rothman, an enclosed box containing a perfect vacuum and two hard balls which are perfect spheres. The box is perfectly insulated so that there are no mechanical vibrations of any sort, no heat differences from point to point, nothing. The only force in the box is that of gravitation.

If one ball is fixed securely to the bottom of the box

and the second ball is allowed to drop downward squarely on the apex of the fixed ball, then, by the classical laws of mechanics, the moving ball will bounce directly upward again, fall down upon the apex once more, bounce directly upward again and so on, for a great many times.

However, the principle of uncertainty would indicate that the ball could not *certainly* hit the exact apex, no matter how carefully it was aimed. And even if it did hit the exact apex, there could be no certainty that it would hit it again the next bounce. Once the position of strike varied from the apex by the tiniest amount, the moving ball would rise in a direction inclined by the tiniest degree to the vertical and would then strike the fixed ball even further from the apex on the next bounce and move up in a direction inclined even more to the vertical, and so on. After ten or twelve bounces, says Rothman, there would be a large probability that the moving ball would miss the fixed ball altogether, no matter how inhumanly carefully it had been aimed in the first place.

A similar situation involves a needle coming to a precise mathematical point. Imagine such a needle balanced on its point in a box free of vibration and heat differences and containing a perfect vacuum. The needle would only remain balanced on the mathematical point if its center of gravity were *exactly* over that point. But, by the principle of uncertainty, the center of gravity could only be within a certain distance of the direct-overheadness. As soon as it departed from direct-overheadness, however slightly, gravity would cause it to depart still further, and it would fall.

In short, the principle of uncertainty makes it impossible to balance a needle on a mathematical point, even under idealized and perfect conditions.

But these are imaginary situations. They involve large-scale objects, yes, but under conditions that cannot, in fact, be set up. Well, then, let's try something else.

As we generally tend to think of absolute zero, it is the temperature at which the energy of motion of atoms and molecules falls to zero. By that view, the vibration of atoms in any substance (all of which are solid in the neighborhood of absolute zero, one would think) slows to nothing and a deathlike and perfect immobility is all that is left.

But that is the view of classical physics, and not that of modern physics. Once the uncertainty principle is accepted, then we can't permit zero energy of motion at any time or under any conditions. If, at absolute zero, atoms were really and truly perfectly at rest, then we would know their velocity to be exactly zero. But we can't ever know velocity exactly. All we can say is that at absolute zero, the energy of atoms is within a certain very close distance to zero velocity, and that the atoms do continue to move, just a little bit.

This small residual "zero point motion" which atoms and molecules retain even at absolute zero represents a *minimum* energy that cannot be removed without violating the inviolable uncertainty principle. For that reason, there can be no temperature lower than absolute zero. Nevertheless, energy content at absolute zero, while a minimum, is *not* zero.

Has this minimum energy any effect that can be observed? Yes, it does. The solid substance whose atoms are most easily jostled apart, and into liquid form, is solid helium. The minimum energy at absolute zero is sufficient for the purpose and the result is that under ordinary conditions, helium remains liquid *even at absolute zero.* Solid helium can only be formed under considerable pressure.

There, then, is a tangible effect of the uncertainty principle under large-scale conditions, and in the real world, not in some unattainable thought experiment.

Is this still too esoteric for you? Is absolute zero and liquid helium too specialized a display of the power of the uncertainty principle to be impressive?

How about this, then? If the uncertainty principle did not exist, neither would the universe as we know it; for the existence of all atoms other than hydrogen depends on the uncertainty principle.

But my space, alas, is used up. . . . Next chapter, please?

Sixteen
BEHIND THE TEACHER'S BACK

In the course of writing these chapters in their original form, I have developed several bad habits. Partly, this is because I have a natural affinity for bad habits and partly because I am given such a free hand that it is hard not to pamper myself.

For instance, when space runs out and I am feeling Puckish, I commit a cliff-hanger and end with an indication that there is more to the story I am telling and that I will reserve the rest for some other time. Afterward, I may write that other column or I may not. It depends on my lordly whim.

A second bad habit is that of constantly referring to my genial audience as "Gentle Readers." The phrase originated, of course, as an indication that the readership was wellborn and possibly of noble descent. In our own egalitarian society, the phrase has lost its aristocratic connotation and one can't help but think of the Gentle Readers as being kindhearted, tender, and sweet. And so they are, so they are, but not always, I'm afraid.

My persistence in the first bad habit has just uncovered an exception to the generalization involved in my second bad habit.

In the original appearance of the previous chapter, I concluded with a cliff-hanger.

Within a matter of days, I received a letter from a Fierce Reader.* Without a gentle word anywhere in it, she slashed away at me for having dared drop the subject of the uncertainty principle in the middle. "Is the continuation coming in the next issue?" she blazed.

I had to reply that it was not; that I hadn't given the matter much thought, and that I had thought that any time in the next year or two would be soon enough. However, since she sounded so savage, I thought I had better change my mind and write the continuation at once.

* A young lady—undoubtedly beautiful.

She replied with the smoldering threat of "You'd better!" I hasten, dear lady, I hasten—

In order not to make things too dull, let's continue the story of the uncertainty principle by talking first of what seems to be a different subject altogether. This different subject we call the Dilemma of the Atomic Nuclei that Shouldn't Be.

The New Zealand–born physicist Ernest Rutherford had conclusively demonstrated the existence of the atomic nucleus by 1911, and for twenty years after that its general structure seemed established. Atomic nuclei were considered to consist of two types of particles, protons and electrons, each of the former possessing a unit positive charge (+1) and each of the latter possessing a unit negative charge (−1). The protons, always present in excess, lent the nucleus a net positive charge.

The one exception to this general rule was the simplest nucleus of all, that of the more common hydrogen isotope. It consisted of a single particle, a proton, and nothing else.

As examples of more complicated situations, the most common oxygen isotope had atomic nuclei made up (it was thought) of 16 protons and 8 electrons for a net charge of +8; the most common iron isotope had nuclei made up of 56 protons and 30 electrons for a net positive charge of +26; the most common uranium isotope had nuclei made up of 238 protons and 146 electrons for a net positive charge of +92; and so on.

This seemed to make sense. The protons in the nucleus, all of them positively charged, repelled each other on the well-known electrical principle that "like charges repel." However, if one placed electrons strategically among the protons, the attraction between protons and electrons (opposite charges attract) would neutralize the repulsions and allow the nucleus to hang together.

The electron could be considered a kind of "nuclear cement" and without it, it would seem, none of the nuclei could exist except that of hydrogen.

Physicists were, however, by no means happy with the atomic nuclei in all respects. By dint of shrewd calculations, they had decided that both protons and electrons had spins that could be characterized by the number +½ or −½. That meant that nuclei made up of even numbers of particles should have overall spins equal to the algebraic sum of an even number of halves, plus or minus. Such a sum would always have to be a whole number, such as 1, 2, or 3.

On the other hand, nuclei made up of an odd number of particles should have spins equal to the algebraic sum of an odd number of halves, and this should always add up to a "half-number" such as 1½, 2½, and 3½.

Unfortunately it didn't work that way. As an example, consider the most common nitrogen isotope. It was made up (so physicists had decided in the Roaring Twenties) of 14 protons and 7 electrons for a total of 21 particles. Since this is an odd number of particles, the nitrogen nucleus ought to have an overall spin equal to a half-number, but it doesn't. Its spin is equal to a whole number.

Something was therefore seriously wrong. Either the nuclei did not have the structure they were thought to have or else the law of conservation of angular momentum was broken.

Physicists did not hesitate in their choice between these two alternatives. They have a peculiar affection for laws of conservation of this or that and would not willingly see one broken. Therefore, they began to turn looks of deep disfavor on the whole proton-electron theory of nuclear structure.

You can imagine, then, the thrill of exultation that swept the world of nuclear physics when, in 1932, the English physicist James Chadwick discovered the neutron, a particle that strongly resembled a proton except that it lacked an electric charge.

Scarcely allowing the neutron's discovery to grow cold, the German physicist Werner Karl Heisenberg (the same who, five years earlier, had enunciated the uncertainty principle) suggested that atomic nuclei were made up of protons and neutrons rather than protons and electrons.

Thus, the nuclei of the most common oxygen isotope would consist of 8 protons and 8 neutrons and its net charge would still be +8, thanks to the protons. (The neutrons, being uncharged, would contribute no charge of their own, and cancel none.) In the same way, the nuclei of the most common iron isotope would be made up of 26 protons and 30 neutrons (net charge, +26), the nuclei of the most common uranium isotope of 92 protons and 146 neutrons (net charge, +92), and so on.

The proton-neutron theory of nuclear structure could explain virtually all the facts of life concerning the nuclei, just as well as the proton-electron theory had been able to. In addition, it also fits the facts of nuclear spin with delightful accuracy. The nitrogen nucleus, for instance, by the new theory, was made up of 7 protons and 7 neutrons, for a total of 14 particles. Now it had an even total of particles and the overall spin could, with sense, be represented by a whole number.

The law of conservation of angular momentum was saved, by George.

There was just one gigantic fly in the ointment. According to the new theory, atomic nuclei other than those of hydrogen ought to be nonexistent.

The electron cement which had been counted upon to keep the protons in happy proximity was gone, and the protons were alone with their own company in the nucleus. (The neutrons were there, but electromagnetically speaking, they didn't count.) The nucleus was full of repulsion, a great deal of repulsion, and nothing but repulsion.

Within the nucleus, two protons are virtually in contact and are therefore separated, center to center, by about a ten-trillionth of a centimeter. The charge on each proton is terribly small by everyday standards, but the distance across which that charge need operate is terribly smaller. The result is that the repulsion between two neighboring protons works out to some 24,000,000 dynes.

This, needless to say, is a simply terrific force to be concentrated between a pair of objects as tiny as protons, and if this were the only force involved, two protons held in such close proximity would remain together for the merest split-instant and then separate at velocities close to that of light. Indeed, in 1932, there was no good way of accounting why two protons should be in such proximity in the first place.

Since in all atomic nuclei, save in that of the more common isotope of hydrogen, two or more nuclei *do* exist in such proximity (with neutrons strewn among them, of course), it turns out that there was nothing, in 1932, to account for the fact that matter, other than hydrogen, existed at all.

Yet even the cleverest and most inevitable scientific reasoning must bow before the presence of even the crudest fact. Matter *did* exist and consequently something was neutralizing and overcoming the repulsion between protons.

The Granger Collection

SIR JAMES CHADWICK

Chadwick was born in Manchester, England, on October 20, 1891. He worked under Rutherford (who discovered the atomic nucleus) at the University of Manchester, and in 1913 went to Germany to do graduate work under Hans Geiger (of the Geiger counter). World War I broke out and Chadwick, as an enemy alien, was interned for the duration.

In 1919, he was back in England, doing research at Cambridge, and was once again under the aegis of Rutherford. At the time, the only two subatomic particles known were the electron and proton. It was thought that the atomic nucleus contained both of these varieties but there were theoretical reasons for being uncomfortable with that notion. Rutherford suspected there might be an uncharged variety of the positively charged proton, but uncharged particles were very difficult to detect and the search for them failed.

Between 1930 and 1932, several physicists noted that protons were being knocked out of nuclei by something that could not be detected. In 1932, Chadwick realized that that "something" were the uncharged particles Rutherford had speculated about. Chadwick called them "neutrons."

This discovery finally put notions of nuclear structure on a firm basis and it was by means of neutron bombardment that uranium fission was discovered. Chadwick received the 1935 Nobel Prize in physics for that.

Chadwick began work toward a nuclear bomb as soon as the fact of uranium fission was announced in late 1939 and before American physicists had been stirred into action. During World War II, he served as head of Great Britain's phase of the nuclear bomb project, spending some time in the United States. He was knighted in 1945, and died in London, on July 24, 1974, in his eighty-third year.

Unfortunately, that "something" had to be another force and there was a shortage of known forces. All the forces known in 1932 were produced by one or the other of just two types of "force-fields."

One of these was the "electromagnetic field," which governed the attractions and repulsions among protons and electrons. It is the presence of this field which keeps atoms from making actual contact, since at close approach the negatively charged electronic layers that fill the outskirts of one atom repel the negatively charged electronic layers that fill the outskirts of the other. Most ordinary forces with which we are acquainted—the pushes and pulls of everyday life—depend on the fact that the atoms of one piece of matter are, at close quarters, repelled by the atoms of another piece.

The only force known in 1932 which was not electromagnetic in nature was that of the "gravitational field," but it was quite clear that gravity could in no way counteract the mighty protonic repulsion within the nucleus. To be sure, protons within the nucleus experienced a gravitational attraction for each other since gravity depends only upon mass, is experienced only as an attraction, and is unaffected by electric charge. Unfortunately, however, the gravitational field is almost unimaginably weak; much less than a trillion-trillion-trillionth as strong as the electromagnetic field.

Something else is needed; some new type of force-field altogether. This would give us a "nuclear force" and if this is to hold nuclei together and make matter (other than hydrogen) possible, it must have certain properties. In the first place, it must be even stronger than the electromagnetic field, at least at close quarters, for it must produce an attraction between protons stronger than the electromagnetic repulsion between them.

Another point— Both electromagnetic and gravitational forces are long range. To be sure, they both weaken with distance, but only as the square of that distance. As a result, gravitational and electromagnetic forces can make themselves felt across vast gaps of space.

The hypothetical force can do no such thing. Within the infra-tiny nucleus it is overpoweringly intense, but it drops off very rapidly with distance; as a high power of that distance, not just the square. At a distance greater than a ten-trillionth of a centimeter (at greater widths than the diameter of an atomic nucleus, for instance) it becomes weaker than the electromagnetic force, and if the distance increases to two or three ten-trillionths of a centimeter it becomes indetectable.

Thus we can explain the fact that protons are strongly attracted at subnuclear distances, but show no signs of attraction if they are anything but cheek-by-jowl.

But we can't just invent a nuclear force of peculiar properties out of thin air. There's something about clear-cut observational evidence which is terribly desirable. To find some, let's go by way of the uncertainty principle. (See, I got there.)

In 1930, at a gathering of physicists at Brussels, Albert Einstein endeavored to show a fallacy in the reasoning that lay behind the uncertainty principle, then three years old. This principle held (as I explained in the previous chapter) that the inherent uncertainty in the determination of position multiplied by the inherent uncertainty in the determination of momentum was equal to not less than about one-sixth of Planck's constant:

$$(\Delta p)\ (\Delta m\dot{v}) = 10^{-27} \qquad \text{(Equation 31)}$$

Einstein showed that if this were so then it was possible to maintain that the same relationship would hold for the product of the inherent uncertainty in the determination of energy content (Δe) and the inherent uncertainty in the determination of time(Δt) so that:

$$(\Delta e)\ (\Delta t) = 10^{-27} \qquad \text{(Equation 32)}$$

He then went on to describe a thought experiment in which both energy and time could be measured simultaneously with unlimited exactness, assuming one had perfect measuring tools. If Einstein were right, the uncertainty principle was out the window.

The Danish physicist Niels Bohr stayed awake that night, and the next day, haggard but triumphant, pointed out a few flaws in Einstein's reasoning and showed that in the thought experiment under discussion, the determination of time would upset the determination of energy and vice versa. Einstein had to admit, reluctantly, that Bohr was right. The uncertainty principle has not been seriously challenged since.

Nevertheless, Einstein's version of the uncertainty principle, in which energy and time are linked, is perfectly correct and it introduces some interesting effects.

Using Einstein's version, imagine that you are measuring the energy content of some system at some instant of time. If your measurement pinpoints the energy content at some mathematical instant of time—over a duration of exactly zero seconds—you can't really measure the energy at all. The uncertainty of energy measurements is then infinite.

If you are content to say that the energy of the system is thus and so over a certain period of time, then you are better off. The longer the period of time, the more exactly you can measure the energy content. Over a period of a ten-trillionth of a second or so, you could, ideally, measure the energy content of a system to a ten-trillionth of an erg or so. Under ordinary conditions, no one would want better than that.

Nevertheless, such a situation introduces a certain limited flexibility into the most important generalization known to science: the law of conservation of energy.

The law states that the energy content of a closed system must remain constant. Energy cannot appear out of nowhere and it cannot disappear into nowhere. However, if you measure the energy content of an atomic nucleus, let us say, over a period of a ten-trillionth of a second, you have determined that energy content, at best, only within a ten-trillionth of an erg. During that ten-trillionth of a second, the energy content can move freely up and down within that limit of a ten-trillionth of an erg, despite the law of conservation of energy. There would be no way of measuring that energy variation and therefore no way of accusing the nucleus of having broken the law.

You may say, of course, that it doesn't matter whether we can detect the violation of the law or not; that the law cannot be violated under any circumstances.

But is that so? Let's take an analogy.

Suppose a schoolboy is strictly forbidden to show any impoliteness to his stern teacher at any time under pain of severe flogging. Suppose further that whenever the teacher turns his back, the boy sticks out his tongue but manages to get it in again before the teacher turns toward him. As nearly as the teacher can tell, the boy is being perfectly polite at all times, and is not breaking the rule.

In other words, a rule which ordinarily can't be broken, can be broken if it is done so over a short enough period. We can make this plain if we reword the rule to make it conform not to an impossible idealism, but to the situation as it truly exists. The rule is not: "A schoolboy must never be impolite to his teacher!" The rule, very obviously, is: "A schoolboy must never be caught being impolite to his teacher." All human rules are of that form. Even a murderer goes unpunished if the existence of the murder goes unsuspected.

Analogously, we must not define the law of conservation of energy as: "The total energy of a system remains constant at all times" but only as: "The total energy of a system remains measurably constant at all times."

What we cannot measure, we cannot expect to insist on controlling by fiat, and the uncertainty principle tells us what we cannot measure. Energy is permitted to vary by a certain fixed amount, and the shorter the time interval over which this variation takes place, the greater the amount of variation permitted.

How does this apply to the nuclear field?

Again we return to Heisenberg. When he suggested the proton-neutron structure of the nucleus he saw very well the difficulty that arose in connection with protonic repulsion. He suggested that force-fields exerted their influences of attraction and repulsion by the exchange of particles between one body and another. In the case of the electromagnetic field, the particle exchanged was the photon (the unit of radiant energy); and in the case of the gravitational field, the particle exchanged was the graviton (a particle which remains as yet hypothetical for it has never been detected).

If there is to be a third force-field, a nuclear one, there must be a third exchange particle.

The Japanese physicist Hideki Yukawa got to work on the properties of this hypothetical nuclear exchange particle.

This exchange particle existed by virtue of the loophole offered by the uncertainty principle. It contained energy, but only the amount permitted by that principle. The shorter the time during which the nuclear exchange particle existed, the more energy it might have, so it was necessary to fix the duration of its existence somehow.

The exchange particle had to exist long enough to get from one proton to its neighbor within the nucleus and back or it would not exist long enough to set up an attractive force between protons. It could not exist much longer than that because then it would last long enough to get outside the nucleus and make the nuclear force felt there—in regions where the nuclear force was never felt. Thus the

The Granger Collection

NIELS HENRIK DAVID BOHR

Bohr, the son of a professor of physiology, was born in Copenhagen, Denmark, on October 7, 1885. He obtained his Ph.D. in 1911, and in 1913 produced the first picture of atomic structure in the light of quantum theory. His picture of electrons capable of existing in certain orbits only was a little hard for older physicists to accept, but it was useful enough to win him the 1922 Nobel Prize in physics.

During the 1920s and 1930s, Bohr headed an institute for atomic physics in Copenhagen that was supported by the Carlsberg brewery. It proved a Mecca for theoretical physicists everywhere so that the city became a major center for advances in atomic studies.

When Hitler came to power in Germany in 1933, Bohr took what action he could on behalf of his colleagues in that terrorized land, doing his best to get Jewish scientists to safety. In 1939, he visited the United States and brought with him the news that the fact of uranium fission was about to be announced. That began the push that ended with the development of the nuclear bomb.

Bohr returned to Denmark and was still there when Hitler's army occupied the land in 1940. His life was in danger thereafter and in 1943, when imprisonment and worse seemed inevitable, he escaped to Sweden and there helped arrange the escape of nearly every Danish Jew from the Nazi grip. He was then flown to England in a tiny plane in which he nearly died from lack of oxygen—and then to the United States where he worked on the nuclear bomb.

His views in favor of international control of the nuclear bomb made him unpopular with Winston Churchill, who wanted to arrest him, but in 1957, he received the first "Atoms for Peace" award. He died in Copenhagen, on November 18, 1962.

time of duration, and therefore the particle's energy content, could be determined within rather fine limits.

Suppose the exchange particle traveled at the velocity of light. It could then cover the distance from one proton to a neighboring proton and back in about 0.000000000000000000000005, or 5×10^{-24} seconds.

If an energy measurement is made over a time interval not less than 5×10^{-24} seconds, the additional energy made available for the briefly existing exchange particle, by the flexibility introduced into the law of conservation of energy by the uncertainty principle, can be determined.

Turning to Einstein's version of the uncertainty principle, the uncertainty in time (Δt) is set equal to 5×10^{-24} and Equation 32 becomes:

$$(\Delta e)\ (5 \times 10^{-24}) = 10^{-27} \quad \text{(Equation 33)}$$

Solving for Δe, we find that it equals 0.0002 ergs. This is the amount of energy that the uncertainty principle makes available for the exchange particle of the nuclear field. It is a tremendous amount of energy for a single particle and it would be difficult to handle as pure energy. It would be more convenient if the energy were, for the most part, packed into the form of mass—which is the most condensed form of energy known. An amount of energy equal to 0.0002 ergs can be packed into a particle with a mass about 250 times that of an electron, with enough left over to give it a velocity nearly that of light.

Yukawa, when he published his theory in 1935, suggested therefore that the nuclear exchange particle have mass (unlike the massless photon and graviton) and that the mass be intermediate between that of the electron on one hand and the proton and neutron on the other. (The proton and neutron are roughly 1,840 times as massive as the electron, and, therefore, something over 7 times as massive as Yukawa's exchange particle.)

Suggesting a nuclear particle of specific properties, is one thing, but some observational evidence was still necessary. Inside the nucleus, the exchange particle comes and goes within the time limit set by the uncertainty principle. This means it cannot be observed under any circumstances. It is a "virtual particle," not a real one.

But suppose energy is added to the nucleus; enough energy to supply the amount required for the exchange particle without having to resort to the flexibility of the uncertainty principle. In that case, might not the exchange particle assume a real existence and condescend to hang around long enough to allow itself to be detected?

The catch was that packing the necessary energy into the small confines of the nucleus isn't easy. In the 1930s, the only possible source of sufficiently concentrated energy were the cosmic rays. In 1936, the American physicist Carl David Anderson, in the course of his cosmic ray studies, found that cosmic rays were indeed occasionally knocking particles out of the nucleus that resembled Yukawa's exchange particle in mass. That mass turned out to be 207 times that of an electron.

Anderson called the particle a "mesotron," from the Greek word *meso* meaning "intermediate," but this was quickly abbreviated to "meson."

Unfortunately, Anderson's meson did not have the properties expected of Yukawa's exchange particle. For one thing, Yukawa's exchange particle had to interact strongly with atomic nuclei, but Anderson's meson did not do so. It virtually ignored the existence of nuclei. The disappointment among physicists was keen.

Then, in 1948, a group of English physicists, headed by Cecil Frank Powell, who were studying cosmic rays in the Bolivian Andes, detected another particle of intermediate mass. The new particle was about 270 times as massive as the electron (about a third more massive than Anderson's particle), and it interacted with nuclei with a most gratifying avidity.

The new particle was also called a meson and it was distinguished from the meson earlier discovered by means of Greek letter prefixes. Anderson's meson was "mu-meson," soon shortened to "muon"; while Powell's meson was a "pi-meson," soon shortened to "pion." It is the pion that is Yukawa's exchange particle.

It is the pion whose existence within the nucleus makes possible the development of a nuclear force of attraction between neighboring protons over a hundred times as intense as the electromagnetic force of repulsion between them. It is the pion therefore that makes the existence of matter, other than hydrogen, possible. And the existence of the pion is itself made possible (behind the teacher's back, so to speak) by the uncertainty principle.

—So be careful how you yearn for certainty.

And what about the muon? If that is not the exchange particle, what is it? That, as it happens, is an interesting question, for in recent years the muon has raised two problems that are possibly the most fascinating that currently face the nuclear physicist. It is not even a meson, really. It is, instead—

But my space, alas, is used up. . . . Next chapter please?

Seventeen
THE LAND OF MU

When I was in my early teens, I found a book in the public library that seemed fascinating. It was *The Lost Continent of Mu* by James Churchward and I took it out exultantly.

The disappointment was keen. I may have been young, but I wasn't so young as not to recognize nonsense. This was my first encounter with the "serious" literature spawned by the Atlantis legend (as opposed to honest science fiction) and I needed no second.

If you want to know more about the Atlantis myth, about Lemuria and Mu, and so on, don't look for it here. I refer you to an amusing and interesting book by a gentleman with the highest rationality-quotient I have ever met: *Lost Continents* by L. Sprague de Camp (Gnome Press, 1954).

For myself, I will go no further into the subject except to say that the Land of Mu was a hypothetical continent that filled the Pacific Ocean and that, like Atlantis, allegedly sank beneath the waves, after having supposedly harbored a high civilization.

Utter bilge, of course, and yet there is a queer coincidence that crops up.

Churchward could not have known, when he wrote his first book on Mu in 1926, that the time would come when the word "Mu" would gain a certain significance in science. This significance rests chiefly in the problems that have arisen in connection with "Mu." These problems, far from being solved, or even moving toward solution, have, in the last quarter century, grown steadily more puzzling and intense until now, in the mid-sixties, the daintiest and most succulent enigmas of the nuclear physicist rest right on "Mu."

There *is* a Land of Mu, in a manner of speaking, and it is far more fascinating and mysterious than the murky lost continent of Churchward's fog-ridden imagination.

Let me tell you about the real Land of Mu, Gentle Readers—

As usual, I will start at the beginning; and the beginning, in this case, is the previous chapter, where I described the efforts made to account for the existence of atomic nuclei despite the strong mutual repulsion of the protons contained in those nuclei. Let me repeat just a little.

To allow the nucleus to exist, the Japanese physicist Hideki Yukawa had found it necessary to postulate the existence of a particle intermediate in mass between the proton and the electron. In 1936, such a particle was found and was promptly named the "mesotron" from the Greek word *mesos* meaning "in the middle" or "intermediate." A syllable was saved by shortening this name to "meson."

Unfortunately, there were two flaws to this happy discovery. The first was that the meson was just a little on the light side. This, however, might not be serious. It might easily turn out that there were factors Yukawa hadn't correctly taken into account in his reasoning.

The second flaw was less easily dismissed. The whole point of the meson was that if it were to serve as the "cement" of the atomic nucleus, it would have to interact very rapidly with the protons and neutrons within that nucleus. It should, indeed, react with a proton, for instance, in not much more than a trillionth of a trillionth of a second. A stream of mesons striking a group of nuclei ought to be gulped up at once.

But this didn't happen. A stream of mesons, shooting forward at great energies, can pass through inches of lead. In doing so, the mesons must carom off vast numbers of atomic nuclei and not be absorbed by any of them.

For a dozen years, that irritated physicists. A particle had been predicted; it was found; and it proved not to be the particle that was predicted. Fortunately, in 1948, a second meson was found, a little more massive than the first, and it *did* react virtually instantaneously with nuclei. The second meson checked out perfectly as Yukawa's predicted particle and physicists have had no cause to question that conclusion since.

It was now necessary to distinguish between the two mesons by name, and one good way was to make use of Greek letter prefixes, a common habit in science.

For instance, the first meson had precedence to the right

to have the prefix "m" for "meson." The Greek letter equivalent of "m" is μ which, in English, is called "mu."* Therefore, the first meson, the one which is *not* Yukawa's particle, was named the "mu-meson" and that is more and more frequently being abbreviated to "muon."

The second meson, which *is* Yukawa's particle, was first discovered among the products of cosmic ray bombardment (the "primary radiation") of the upper atmosphere. It therefore should have the initial "p" for "primary." The Greek letter equivalent of "p" is π which, in English, is called "pi." Therefore, Yukawa's particle became the "pi-meson" or "pion."

Suppose, now, we set the mass of the electron at 1, and consider the masses of the two mesons and of the proton and neutron, as in Table 9.

TABLE 9—The Masses of Subatomic Particles

electron	1
muon	206.77
pion	273.2
proton	1836.12
neutron	1838.65

If we look at this group of particles, we can say that the atom is made up of a nucleus, containing protons and neutrons held together by pions, and that outside that nucleus are to be found electrons. Here are four different particles, all essential to atomic structure.

But that leaves the muon; the mu-meson; the particle which we can name, if we are feeling romantic enough, the physicists' "Land of Mu." What does it do? What function does it serve?

Do you know that it is just about thirty years now since the muon was discovered and physicists still don't know what it does or what function it serves.†

This puzzle—the first of the Land of Mu—is not as acute as it might be. In the 1950s and 1960s, other particles were

* The continent of Mu, concerning which Churchward wrote, has no connection at all with the Greek letter mu, but I am building this article on the coincidence all the same.

† This article first appeared in October 1965. Nine more years have passed and physicists *still* don't know.

discovered by the dozens. There then arose the question of accounting for the existence and function of a large number of particles and not the muon only.

There is the feeling now that what is needed is a general theory covering subatomic particles as a whole. Indeed such general theories are now being advanced, but this chapter is not the place to discuss them.

However, other problems arose in connection with the muon that apply strictly to the muon and to no other particle; and concerning which physicists haven't yet the foggiest beginning of an answer.

Suppose we compare the muon and the electron, for instance.

Item 1: The electron carries a negative electric charge, arbitrarily set equal to unit-size, so that its charge is described as −1. It has a positively charged twin, the positron (charge, +1). There is, however, no such thing as a "neutral electron" (charge, 0).

Should there be? Well, there is a "neutral proton" (charge, 0), which we call a neutron, which is a trifle more massive than a proton (charge, +1) or its twin, the antiproton (charge, −1). There is a neutral pion (charge, 0), which is a trifle less massive than the positive pion (charge, +1) or its twin, the negative pion (charge, −1).

Nevertheless, although the proton and pion can exist in uncharged form, the electron apparently cannot. At least a neutral electron has never been detected and there is no theoretical reason to suspect that it might exist.

Now the muon. There is a negative muon (charge, −1) and a positive muon (charge, +1) but there is no neutral muon.

Indeed, in May 1965 (three days ago, as I write this), the results of an experiment were announced in which streams of muons and of electrons were bounced off protons. From the nature of the scattering one could calculate the volume over which the electric charge of a muon and of an electron were spread. No difference could be detected in the two particles. We can conclude then that in terms of nature and distribution of charge, the muon and electron are indistinguishable.

Item 2: Subatomic particles have a property which can be most easily described as spin about an axis. The spin of an electron (or positron) can be expressed as either +½

The Granger Collection

HIDEKI YUKAWA

Yukawa was born in Kyoto, Japan, on January 23, 1907, and was educated at Kyoto University, where his father was a professor of geology. He graduated in 1929, then went on to do graduate work at Osaka University. He got his Ph.D. there in 1938, while already serving on the faculty.

It was while he was still only a graduate student that he

began to work on the manner in which the nuclear force was expressed. He published his meson theory when he was only twenty-eight years old, and might have obtained the Nobel Prize for it earlier, were it not for the turmoil of the late 1930s and early 1940s.

He worked in Japan throughout World War II, and it was one of the signs that the world was returning to normal that, in 1948, three years after the atomic bombs over Hiroshima and Nagasaki had ended World War II, J. Robert Oppenheimer invited Yukawa to visit the Institute for Advanced Study at Princeton. He remained in the United States till 1953, lecturing at Columbia University, then returned to Kyoto.

While he was in the United States, he obtained the Nobel Prize in physics in 1949, the first Japanese to win a Nobel award.

The meson theory is not Yukawa's only theoretical accomplishment. In 1936, he had predicted that a nucleus would absorb one of the innermost of the circling electrons and that this would be equivalent to emitting a positron. Since the innermost electrons belong to the "K shell," this process is termed "K capture." Yukawa's prediction was verified in 1938.

or $-\frac{1}{2}$. The spin of a muon (either negative or positive) can also be expressed as either $+\frac{1}{2}$ or $-\frac{1}{2}$. In terms of spin, then, the muon and electron are indistinguishable.

Item 3: The spin of the electron (or positron) sets up a magnetic field whose strength can be expressed as 1.001160 Bohr magnetons. The spin of the muon (either negative or positive) sets up a magnetic field of strength equal to 1.001162 Bohr magnetons. The difference is quite unimportant and in terms of magnetism, too, then, the muon and electron are just about indistinguishable.

Item 4: The electron and muon can take part (or fail to take part) in certain interactions with other particles. For instance:

a) A pion is an unstable particle which, left to itself, breaks down in a couple of hundredths of a microsecond.†† When it breaks down, it may produce a muon and a neu-

†† A microsecond is equal to a millionth of a second, or 10^{-6} seconds.

trino; or it may produce an electron and a neutrino.* In brief, either an electron or a negative muon can be produced by the breakdown of a negative pion; either a positron or a positive muon can be produced by the breakdown of a positive pion.

b) An electron has very little tendency to interact with atomic nuclei; it tends to remain outside the nucleus. A muon has very little tendency to interact with atomic nuclei; it tends to remain outside the nucleus. In fact, negative muons can replace electrons in their orbits about nuclei, producing what are called "mesonic atoms." No doubt, positive muons could replace positrons in their orbits about the nuclei of anti-matter to produce "anti-mesonic atoms."

c) It is possible for an electron and its positively charged twin, the positron, to circle each other for short intervals of time, making up a neutral system called "positronium." It is also possible for the electron to circle a positive muon in place of the positron, to form "muonium." Undoubtedly, a positron can circle a negative muon to form "anti-muonium," and a negative muon can circle a positive muon to form something for which, as far as I know, no name has yet been coined ("mumuonium"?).

We can conclude then that as far as the nature of particle interactions in which electrons and muons can take part is concerned, the two are indistinguishable.

In fact, as physicists examined the muon more and more thoroughly through the 1950s, it began to dawn upon them that the muon and electron were identical.

Well, almost identical. There remained two important points of difference.

The first involves the matter of stability. The electron and positron are each stable. An electron or a positron, alone in the universe, will never (as far as we know) alter its nature and become anything else.

The muon, however, is unstable. Even if alone in the universe, it will break down, after an average lifetime of 2.2 microseconds. The negative muon will break down to an electron and a couple of neutrinos, while a positive muon will break down to a positron and a couple of neutrinos.

This certainly seems to be a tremendous distinction between electron and muon, but, oddly enough, it isn't. A microsecond is a brief interval of time on the human scale

* I'll have something to say about neutrinos later in the article.

but not on the subatomic scale. On the subatomic scale there are reactions and breakdowns that take place in 10^{-23} seconds. The pion interactions that serve to hold an atomic nucleus together take place in such an interval of time.

If we take 10^{-23} seconds as a normal interval of time on the subatomic scale; short but not extraordinarily short; we might compare it to 1 second on the human scale. In that case 10^{-6} seconds, which is a hundred quadrillion times as long as 10^{-23} seconds, would be the equivalent of three billion years on the human scale. If we consider the muon from the viewpoint of the subatomic world, it lasts billions of years and surely that is "practically forever."

The difference between an electron that lasts forever and a muon that lasts practically forever is not great enough to bother a physicist.

You yourself may, however, feel more hardheaded than a physicist. "Practically forever" is not "forever" you may exclaim, and six billion years is not eternity.

In that case, look at it another way. When subatomic particles break down, the tendency is to form a lighter particle, provided none of the rules of the game are broken. Thus, when a negative muon breaks down it forms the negatively charged, but lighter electron. The mass decreases but the electric charge remains, for the rules of the game decree that electric charge may not vanish.

The electron, however, cannot break down for there is no particle less massive than itself that carries an electric charge. The charge must remain and therefore the electron, willy-nilly, must remain.

The positron remains stable, too, because there is no particle less massive than itself that carries a positive electric charge.

(Of course, an electron and a positron can undergo mutual annihilation, for the negative charge of the first cancels the positive charge of the second. The total charge of the two particles before annihilation is $+1$ plus -1, or 0, so that no *net* charge is destroyed.)

In short, the fact that the electron lasts forever and the negative muon lasts only practically forever is entirely a matter of the difference in mass. We can ignore the difference in stability, then, as a purely derivative matter and pass on to the difference in mass, which seems essential.

The negative muon is 206.77 times as massive as the electron; the positive muon is 206.77 times as massive as the positron. Up through 1962, all other differences between these two sets of particles seemed to stem directly from this difference in mass.

The case of stability versus instability is a case in point. Here is another. When a meson replaces an electron in its orbit about an atomic nucleus, to form a mesonic atom, the meson must have the same angular momentum as the electron. The angular momentum increases with mass and also with distance from the center of rotation. Since the meson is over 200 times as massive as the electron, it must make up for that increased mass by decreasing its distance from the center of rotation (the nucleus) correspondingly. In very massive atoms, which ordinarily draw their innermost electrons into close quarters indeed, the meson, drawn closer still, actually circles within the nucleus's outer perimeter. The fact that the muon can circle freely within the nucleus shows how small the tendency is for the muon to interact with protons or neutrons. (It also raises puzzling questions as to the nature of the inner structure of the nucleus.)

In addition, the electrons of an atom, in shifting from energy level to energy level, typically emit or absorb photons of visible light. The more massive meson shifts over energy gaps that are correspondingly greater. The protons such atoms emit or absorb are over 200 times as energetic as those of visible light and are in the x-ray region.

The peculiar structure of a mesonic atom and its ability to emit or absorb x rays can thus be seen to be merely another consequence of the great mass of the muon.

Here is still another example. A pion, in breaking down, can form either a muon or an electron. One might think that, if the muon and electron were exactly alike, each ought to form with equal probability and as many electrons as muons ought to be formed.

This, however, is not so. For every electron formed, seven thousand muons are formed. Why is that?

According to the theory of such interactions, the probability with which a muon or an electron is formed depends on how far short of the speed of light the speed of the particle formed happens to be. The electron is a very light particle and at the moment of formation is fired out at almost the speed of light. Its speed is only slightly less than the speed of light and the probability of its formation is correspondingly small.

The muon, however, is over two hundred times as massive as the electron and is therefore considerably more sluggish. Its velocity, when formed, is quite a bit less than the speed of light and the probability of its formation rises correspondingly. The difference between the quantity of muons formed in pion breakdown and the quantity of electrons again boils down to a consequence of the difference in mass.

Physicists have therefore taken to looking at the negative muon as nothing more than a "massive electron" and at the positive muon as a "massive positron."

And that is the second puzzle from the physicists' Land of Mu. Why should the muon be so much more massive than the electron? And just 206.77 times as massive, no more or less? No one knows.

For that matter, why should this huge difference in mass make so little difference in respect to charge, spin, magnetic field, and type of interactions undergone? No one knows.

And even yet we are not done. I've saved the most recent and most tantalizing puzzle for last.

There are certain massless, chargeless particles called neutrinos which have, as their twins, anti-neutrinos. (They are opposite in terms of the direction of their magnetic fields.) These particles are particularly associated with electrons and positrons.

When an electron is formed in the course of a particle breakdown, an anti-neutrino is formed along with it. When a positron is formed, a neutrino is formed along with it.

When a negative muon is formed in the course of a particle breakdown, an anti-neutrino is formed along with it, too. And, of course, when a positive muon is formed, there comes the neutrino.

At first it was thought that since the muon was more massive than the electron, the neutrino produced along with muons ought to be more massive than those produced along with electrons. Consequently, physicists distinguished among them by speaking of a "neutretto" as being associated with muons.

However, the closer they looked at the "neutretto," the less massive it seemed to be until, finally, they decided the "neutretto" was massless.

But the only difference between muon and electron was mass, and if that difference was wiped out between the neutretto of one and the neutrino of the other, there was no difference left in the latter case.

Both were neutrinos (and anti-neutrinos). Physicists decided then that the muon's neutrino (and anti-neutrino) and the electron's neutrino (and anti-neutrino) were the same particle in every respect. This seemed but another example of how the electron and muon could not be distinguished except by mass and mass-derived properties.

Yet one problem remained. When a negative muon broke down, it formed an electron and *two* neutrinos. From theoretical considerations, it was necessary to consider one of those neutrinos a neutrino and the other an anti-neutrino. But a neutrino and an anti-neutrino should be able to annihilate each other and leave nothing behind but electromagnetic radiation. In that case, a negative muon should break down to form an electron as the only particle. And a positive muon should break down to form a positron as the only particle. At least, this should be observed once in a while.

However, it was *never* observed. The neutrino and anti-neutrino were always formed in the breakdown of either the negative or positive muon and never annihilated each other. Physicists began to wonder if perhaps the neutrino and anti-neutrino didn't annihilate each other because they *couldn't* annihilate each other. Perhaps there were two kinds of neutrinos and two kinds of anti-neutrinos after all, one set associated with electrons and one with muons, and perhaps the neutrino of one set could not annihilate the anti-neutrino of the other set.

Could it be then that a negative muon broke down to form 1) an electron, 2) an electron-anti-neutrino, and 3) a muon-neutrino? And could a positive muon break down to form 1) a positron, 2) an electron-neutrino, and 3) a muon-anti-neutrino?

If so, that would explain the facts of muon breakdown. Nevertheless, the possibility of two kinds of neutrinos seemed too much to swallow without additional evidence.

In 1962, therefore, at Brookhaven, Long Island, a "two-neutrino experiment" was set up and carried out. High-energy protons were smashed into a beryllium target under conditions that formed a high-energy stream of positive and negative pions. These broke down almost at once to positive and negative muons. The positive muons, when formed, were accompanied by muon-neutrinos, while the negative muons, when formed, were accompanied by muon-anti-neutrinos.

Before the muons had a chance to break down, the stream struck a wall of armor plate about forty-five feet thick. All pions and muons were stopped, but the muon-neutrinos and muon-anti-neutrinos went right on through. (Neutrinos can pass through light-years of solid matter without being stopped.)

On the other side of the armor plate, the neutrinos and anti-neutrinos had a chance to interact with particles. They did this only very rarely, but every once in a while one of them did. A neutrino (charge, 0) could strike a neutron (charge, 0), for instance, and form a proton (charge, +1).

The rules of the game, however, say that you can't form a positive electric charge out of nothing. If one is formed, then a particle with a negative electric charge must be formed simultaneously so as to keep the total charge zero.

Therefore when a neutrino and neutron combine to form a proton, they must also form either an electron or a negative muon, since either of them can supply the necessary charge of −1 to balance the proton's +1.

All the neutrinos in question were formed along with muons and are therefore muon-neutrinos. If the muon-neutrinos were indeed different from electron-neutrinos, then only muons should be formed. If the muon-neutrinos were identical with electron-neutrinos, then the neutrinos could be considered as associated with either, and some of each ought to be formed. They might not be formed in equal numbers, for the mass difference might apply, but they both ought to be formed.

However, in the first experiment and in all that followed, muons, and muons only, have been formed. Electrons have never been observed.

The conclusion is that there are indeed two different neutrino/anti-neutrino pairs; one pair associated with electrons and positrons, and one pair associated with negative and positive muons.

And there is the third puzzle from the physicists' Land of Mu. What in blazes is the difference between a muon-neutrino and an electron-neutrino?

Both are massless. Both are chargeless. Both have a spin of ½. Put each in isolation and the physicist cannot imagine how to distinguish one from the other.

But the neutron can tell. It will interact with one of them

to form a proton and a muon, and with the other to form a proton and an electron. And how does the neutron tell the two neutrinos apart when we can't?

No one knows.

There you have the physicists' Land of Mu. Compare all this with Churchward's Land of Mu, which could produce nothing more than the imaginary sinking of a mythical continent, and tell me where the true romance lies.

INDEX